Demolishing Darwin

Dismantling Materialism

Why Darwin's Theory of Evolution is No Longer
Good Science

Daurie Laurence

To all who keep their minds open in the scientific quest for what may really be true.

With grateful acknowledgements to those writers and publishers whose work is quoted or drawn from here: John Lennox, Matti Leisola, Valtaoja, William Dembski, Jonathan Witt, Michael Denton, Michael Behe, Mark Eastman, Chuck Missler, Dr Stephen Meyer, Dr. Antony Latham, Jonathan Wells, Chris Carter, Art Battson, Charles Darwin, Michael Flannery, John Davidson, Luther Sutherland, Peter Tompkins, James Perloff, Dean Radin, Rupert Sheldrake, P. M. Atwater, Lyn Margulis, Michael Polanyi, Colin Patterson, V. J. Torley, Professor James Tour, Neal Grossman, Sandie Gustus, Steve Taylor, Michael Egnor, Casey Luskin, Tom Bethell, the Journal of the Discovery Institute, *Evolution News,* Dr Bradley Nelson, Anne Horne, Dr Mary Neal, Anita Moorjani and many others, too numerous to mention, researching into the non-physical aspects of reality, including paranormal phenomena, over the last 150 years.

First published in Great Britain in 2019 ISBN 978-1-78963-080-0

Contents

Preface

Do you believe in evolution or is that an odd question? Evolution is not something to believe in but just a fact, like electricity or that our magical Earth is round not flat? This is what I used to think. But, after a time in my teens as a very firm fan of Darwin and his theory, I began to have certain doubts and, eventually, to discover that his famous ideas were not quite as well-proven as I had been led to believe. In fact, I was very interested to discover that Darwin's ideas about Life were as much driven by his *philosophy* as by the evidence.

This book is my attempt to share with you my doubts and my discoveries, as I researched deeper into this vital subject. One of my surprising finds was that Darwin himself was, at times, prone to profound doubts about his own claims. Yet, ultimately, he disregarded them and put his philosophy and his belief system ahead of all the evidence which tended to falsify his theory.

Darwin, controversially for his time, inclined to *materialism,* the belief that only the physical universe is real, that there are no subtle *spiritual* causes, *no super-intelligent, Creative Energy or Energies behind things,* of the like to which humanity gives its traditional religious names. No, Life and Mind were not, ultimately, supernatural phenomena, as, until then, most had believed. They were probably purely physical things which could be explained in purely physical terms. No need to invoke anything spiritual or supernatural.

While it was radical at the time, Darwin's once controversial philosophy is now mainstream and is considered by many to be the only possible *scientific* view. Science and Darwin, many now believe, have shown Life to be a mere accident. All talk of the supernatural or the spiritual just wishful thinking.

However, for a theory to be scientific, it must be testable. So, if Darwin's theory *is* tested, and if it is found that life could *never* arise by chance, and if it is found that, despite all the claims, the evidence does not, after all, bear out his ideas, then something *more* than mere luck must be involved in Life's arisings. If so, much of what we have thought about the history of Life since Darwin first published will be called into question and there will need to be a massive and, for many, a mind wrenching rethink about our true origins.

This book's cover illustrates some of the key ideas we'll consider. Did particles evolve into people by pure chance? Non-living molecules into single cells, single cells into fish, fish into amphibians, dinosaurs into birds, bears into wales, small mammals into people? Are we spiritual beings having physical experiences? Or are we cosmic accidents as Darwin believed?

Introduction

Existence is a mystery. That there is anything at all, rather than nothing at all, is astonishing. She or he who says otherwise isn't paying attention. But is this curious cosmic matrix in which we live and move and have our being, mindless or meaningful?

Is it, fundamentally, random, or is there a behind-the-scenes intelligence and a purpose to it all? And, can we work out, by various means, including *scientific* means, which of these two most basic ideas about reality is the more likely? Does Life arise due to intelligent Causes, which just mysteriously are, or due to mindless Causes, which – just as mysteriously – just are?

Until a century or so ago, almost everyone, rich and poor, well known or obscure, held to the first view. Life and Mind had intelligent causes. They were fascinating miracles of the subtle Unseen Causes of things. Beliefs poetically expressed in the world's traditional creation stories. Visible nature derived, ultimately, from mysterious, physically unseen, spiritual sources.

Then, as modern science developed, and became increasingly effective at solving all kinds of mysteries, some thinkers, like Darwin, began to question the world's traditional, religious creation stories, at least as literal truth, and they started to look for more scientific and, sometimes, even *wholly materialistic* explanations for the facts and questions of Life.

Facts like? Living things. What breathed Life into them? Did they have intelligent causes, or mindless? Magical Birth. Cosmic Miracle or Cosmic accident? Mysterious Death. Ultimate ending or dimension-shifting transition?

What about Mind and Feelings, Love and Desire, Thinking and Creativity? Were these curious Soul qualities and abilities mere accidents of a matter that was, itself, mindless and insentient, or something more? Nature's Laws? Plants and animals, vertebrate and invertebrate, crawling and flying, water breathing and air breathing? The mysterious fossils. How had the species come to be?

Did they arise all at once, or gradually? From common ancestor microbe to spreading tree, as Darwin imagined? Or, were they more like a woodland, consisting in many different types of tree, planted separately, and at different times? Were they products of *an underlying intelligence* – to which humanity gave its traditional religious names? Or were they produced by wholly *materialistic and coincidental* forces, not supernatural forces at all?

Deep questions, and their answers so key to our understandings of *who and what* we really are. They were not, though, the kind of questions I gave a lot of thought to when, as a teenager, I first met Darwin's ideas, enjoyed reading of his famous voyage on the *Beagle,* of his simple, steam age theory of evolution, which purported to answer some of these questions, and consequently lost all my prior beliefs in the subtle, hidden causes of things.

As a not very deeply thinking teen, I swallowed his materialist evolutionary scheme – by random variations away from random common ancestor microbe into millions of new species – without too much reflection and concluded that he had found an explanation for Life which made any belief in the supernatural unnecessary.

I became a convinced materialist and a passionate atheist, a kind of young Richard Dawkins, and I began to consider all those who still held spiritual beliefs naive or superstitious. Yet, just a few years later, I began to doubt.

There was so much Darwin's ideas didn't explain and I began to realize that there might more be to reality than he and other famous materialists realized.

Yet, for many today, to question Darwin is considered unreasonable. Why? Because the evidence for his theory is said to be overwhelming and any notions of a meaningful creation, or of a purposeful unfoldment of Life through time, are held to be outdated and unscientific. Darwin, it is held, demonstrated that *wholly unguided processes* gave rise to all things, including, eventually, to humanity with all its abilities – creative, artistic and scientific.

It is, therefore, only the intellectually feeble or the emotionally fragile who will continue to cling to the ancient notion that there is anything meaningful to the ways in which living things are put together. No, Darwin showed, just give it enough time and the random forces of chaos plus the precise, but equally mindless, laws of physics and chemistry could, and did, give rise to all.

This is why, given the widespread adoption of his ideas, it can come as quite a surprise to discover that, not just a few but, almost *all* the key predictions of his famous theory have *not,* after all, been confirmed by the real-world evidence of the fossils, genetics, embryology and biochemistry.

'Despite all the claims, published in so many books and studied in so many schools and universities, all over the world, when looked at just a little more carefully, Darwin's famous ideas, and his *materialist* philosophy driving them, fail to accommodate the data. This claim may seem outlandish, even ridiculous to some, but to make it, and to substantiate it, is the purpose of this book.

Yet, so far, those who question Darwin or his philosophy tend to be sidelined or ridiculed. And, for a while, I was one of the mockers. I regarded all who didn't believe in his materialist gospel irrational. Then, in my twenties, I took up meditation, just for relaxation, with no spiritual aims in view. Yet, as I read further on the subject, I found myself becoming more open once again to the classical idea that there might, after all, be more to reality than Darwin and modern science had led us to believe. I also read some of the *modern – non-religious – evidence* for near death and other supernatural experiences and this, also, led me to question my materialist convictions.

At the time of my teenage conversion to materialism, via Darwinism, which seemed so much more 'scientific' to me than the religion of my childhood, I didn't realize that just because the world's traditional creation stories are not literal or scientific descriptions of natural history, it doesn't mean they contain no truth at all. Just because the simplistic creation story which goes,

'In the beginning were the wonderful Mr Rolls and Mr Royce and they said,
'Let there be Rolls Royces!' and, 'Lo, there were, and it was good!'

is not an engineer's description of the famous cars' genesis, it is perfectly true, in its own poetic way. Likewise for,

'In the beginning God created the heaven and the earth. ... And God said, Let there be light: and there was light. And God saw the light, that it was good: and God divided the light from the darkness. . . .' Genesis.

These words, while not science, may, in their own way, contain some truth. On the other hand, *scientific* descriptions of the universe are not intended to be poetic or religious. Nor do they give us any clues as to the meanings and purposes of existence – if there are any.

Describing the Big Bang and later events in scientific terms, clever as it is, can simply amount to *a listing of an order of physical events,* merely a history of the universe and some of its transformations through time.

But such a description of the myriad transformations of materials and mechanisms through time does not, necessarily, amount to a full explanation.

'At 0 seconds the Big Bang happened and the fundamental forces were unified. At $(0 + a)$ seconds gravity separated, leaving the other forces unified. At $(0 + b)$ seconds the stars appeared, at $(0 + c)$ seconds Life on Earth appeared etc.'

Similarly, someone could describe a fully automated car factory:

'At $(0 + a)$ seconds the chassis was made. At $(0 + b)$ seconds the engine was mounted. At $(0 + c)$ seconds the wheels were added, etc.'

Neither of these descriptions of a series of events explains *why* there should be a Big Bang, or an automated factory, followed, in the case of the universe, by the arising of *incredibly* precise physical laws, smart-materials chemicals or complex, DNA CODED Life. They don't tell us of the deeper causes of things.

Although, according to the modern materialist view, Darwin's view, nowadays a very fashionable view, there are no deeper causes. However, is materialism *true?* Has science shown it to be? Was Darwin right to believe that *physical* nature is the only kind? Life and Mind were just amazing cosmic flukes? We were not, not after all, the meaningful products of meaningful causes but merely the random children of a soulless universe which didn't care less about us, because it couldn't care?

Science can help us try to answer some of these questions and the answers it gives us on origins are among the most important of all. If we really are in a wholly material universe, with no *spiritual dimensions,* one consisting solely in physical matter, just mindless rocks and stones, we are truly alone.

If this bleak scenario is true, our *bright* minds must have emerged from the dull waters and mindless rocks of our planet by pure coincidence.

On the other hand, most of humanity, and most of its greatest minds, have always believed there is more to us than mindless chance + mindless laws + mindless matter, that there are nested hierarchies of highly creative *spiritual laws, forces and agencies* within or behind, above or beyond obvious physical reality, despite its many random aspects, all emerging from a Final Source and Uncaused Cause of unimaginable power, beauty, creativity and intelligence.

'Oh, you mean God?' Well, some call it that. Others give it different names: *the Tao or Brahman, Allah or Buddha Nature, the Great Spirit or the Great Mystery.* But, while, for some, these words are interchangeable, for others, God is not the same as the Tao and Brahman is not the same as Allah because of their differing beliefs around these terms.

But the one thing all spiritually minded people do have in common is their belief – contra Darwin and my teenage self – that there is more to reality than mindless dust + mindless laws + mindless chance. And, for our purposes, the differing shades of opinion among the world's believers, as to the precise characteristics of the *ultimate Spiritual Nature* they believe in, are not important because we'll be looking at a far more basic division, that between materialism and spiritualism, (sometimes also referred to as *idealism).*

The central belief of materialism, (sometimes referred to as *naturalism),* is that only *physical* nature is real. If this is true, Life and Mind must just be very lucky coincidences of physics and chemistry, there's no other option.

Materialism versus *spiritualism,* naturalism versus *super naturalism,* 'deadism' versus *vitalism.* Which of these two most basic and most opposed of all possible world views better accords with the data? Which is the more *scientific?* Is it good science to affirm that only the physical is truly real? But

Darwin, and other scientists, have discovered that, due to a long and incredibly unlikely series of chemical coincidences, parts of it became alive, then fish shaped, eagle shaped and people shaped, all for no reason or point at all?

Is it good science to hypothesize that this *quiet, intelligent, Beingness, which looks out through our eyes, this quiet, self-aware, self-Awareness,* is just a cosmic fluke? And every single spiritual or mystical experience anyone has ever had, which point to the *multi-dimensional and the spiritual,* within or behind the obvious physical worlds, are either hallucinations or hoaxes?

Or, is it more reasonable to hypothesize that there is more to us than pure chance? For the world's materialists, like Darwin, it is the 'mindless-chance-creates-everything' hypothesis which is the best view and we are, as Richard Dawkins, Darwin's most famous modern fan, puts it,

> 'In a universe of electrons and selfish genes, blind physical forces and genetic replication, some people are going to get hurt, other people are going to get lucky, and you won't find any rhyme or reason in it, nor any justice. The universe that we observe has precisely the properties we should expect if there is, at bottom, no design, no purpose, no evil, no good, nothing but pitiless indifference.'[1]

In such a universe there are no spiritual dimensions and Darwin's theory, or something similar, is the only possibility. On the other hand, if there is *'something more'* how does it give rise to reality? Is it instantly? Or is it by a mix of more rapid creation at certain times and slower evolution at others?

Our purpose here is to look at some of these issues and to question the materialist philosophy which currently rules mainstream science. Because, in profound disagreement with my teenage self, I now believe materialism to be a mistaken belief system, corrosive to science and to many other fields as well.

While once a passionate materialist myself, I gradually became skeptical that thinkers like Darwin or Dawkins really had solved the mysteries of Life.

But what qualifies me, or anyone else, to express any doubts about Darwin and his philosophy? Is the growing suspicion that his theory doesn't stack up enough? For those on the other side of these arguments it is all very easy to say, without giving it too much thought, just as I once failed to do,

> 'Life is just a wonderful accident. NON-LIVING chemicals *'naturally' evolved* into (stunningly) complex DNA CODED, *LIFE!* "How?" Well, it's simply a matter of time, probability and natural law. There were pre-life, chemical-soup seas, flashes of lightning and, 'Hey presto, Life!' It was a purely 'natural' process. *No intelligence was involved.* Simple *single* cells

---

[1] Dawkins, R. *River Out of Eden: A Darwinian View of Life.*

[which are not simple at all but astonishingly complex!] arose by *pure chance.* These early cells then evolved, *also by chance,* into more complex *multi*-celled worms, trilobites and fish.

'The water-breathing fish evolved into air-breathing amphibians, then dinosaurs, the dinosaurs into birds, with *wholly different lungs* again, land mammals, like bears, evolved into deep-sea diving whales, caterpillars into butterflies, shrews into bats and ape-like animals evolved into people, all for no reason or purpose at all. *Intelligence played no part at any stage.'*

It is ironic that so many people, today, both scientists and lay people, have replaced an outdated, medieval, religious-based creationism with lazy claims like these. Without too much thought, they breezily tell us that this or that stunningly complex living thing simply "evolved," no intelligence was involved – without bothering to see if their claims have any genuinely scientific or rational basis. That these kinds of claims can easily be shown to be false is generally ignored. However, as we will soon see, so glaring are the problems with Darwin's ideas that it is really very easy, even for non-scientists, to pick up on them.

Yet there were plenty of scientists, even in Darwin's day, who were skeptical of his predictions and claims. That he seemingly succeeded, and they were sidelined, says as much about our changing beliefs over the last couple of centuries as it does about our science and its accuracy. Here, you will hear from some of them and their successors. They deserve to be heard.

But if all this is so obvious that anyone can see the flaws in Darwin's ideas, why do more people not question them? I think it's partly fashion and partly schooling, where the young are often led to believe that,

"Darwin, the greatest scientist of all time, solved the riddles of living things. He, pretty much, explained it all. Only the details remain to be filled in."

Then, student days left behind, many of us are just too busy getting on with our lives to give these matters much further thought.

Others feel let down by their childhood religions, or by life itself, and, in their disillusionment, they adopt Darwinism and materialism as their basic creeds. And, for many practicing scientists these days, it can be risky to question Darwin or to show any interest in the supernatural in general.

Darwin, many now believe, showed *materialism* to be the better view and all notions of the supernatural to be delusions. It was, above all, the adoption of *his* ideas by influential intellectuals, in the 19th and 20th centuries, which contributed to the great collapse of religion and spirituality throughout the west and in many other parts of the world as well. God was well and truly dead. This is why, if Darwin was mistaken, it presents such a profound challenge to materialism: a narrow, reality-shrinking belief system which, in the name of respectable, responsible science, tries to explain away nature's elegant laws, our complex bodies and our *bright* minds as mere chemical coincidences, declares all talk of God and Spirit, of Meaning and Purpose, meaningless and asks us to be 'intellectually satisfied' with its dubious, unproven fare.

This is why, if Darwin was wrong, and if he can be shown to be, it will allow *magic, Soul and meaning* to be restored to our world conceptions.

This is not just some abstract or trivial thing. Every year, countless children and young people, with at least some belief in the spiritual, go to schools and colleges all over the world. Then, just as I, and perhaps you, once did, they learn about Darwin's ideas and, in the name of science, this most respected and most trusted of disciplines, they begin to doubt whether there is anything more to reality than pointless matter and pointless chance. Not only is this depressing, and it can make life seem meaningless, what if it's just wrong?

Eventually, we all pass through the fascinating mystery we call death. Does it not matter, then, whether our living and our dying are part of something meaningful and ongoing or, whether, cosmically, they mean nothing at all?

That Life evolved by wholly mindless means is the central claim of materialism and this view is, currently, the mainstream scientific position. Life is just an accident, which evolved by accident, and Darwin proved it.

But for this claim to be genuinely *scientific* it must be capable of testing. Otherwise it's just *faith and ideology*. Well, it can be tested and it has been tested, very thoroughly by now. What is not yet widely admitted, though, is that it has failed those tests, not just once, or twice, but *repeatedly.*

The evidence, it is now clear, does *not* support Darwin's once intriguing seeming ideas. Yet, for reasons of philosophy and professional prestige, many of those most influential in science today are, so far, unwilling to admit this. This is partly because the debates around the origins of Life and the true nature of reality can be so charged.

For example, many now believe that science and religion are fundamentally in conflict, and that either we have to take the world's traditional creation stories as literal truth or we must disbelieve altogether. Either *religious* fundamentalists are right or *atheist–materialist* fundamentalists, like Richard Dawkins, are right. There's no middle ground. This is not true and it's a false choice.

The middle ground being that the *super* natural is real, that, as we'll briefly overview, there is good quality, modern – *non-religious* – evidence for it and the world's religions are not, necessarily, the only possible sources of information about the *non-physical* elements of reality. The world's religions may continue to contribute to our understandings of the spiritual dimensions of existence but they are not the *only* nor, necessarily, the most reliable means by which we can gain some modern knowledge of these subtle things.

A friend once mentioned to me his admiration for the physicist Richard Feynman and for Feynman's comment that he'd "rather have questions [in science] that can't be answered than answers that can't be questioned."

Unfortunately, in our often troubled and confused times, Darwin's theory, and the materialist beliefs behind it, have, for many, become answers to Life's mysteries which cannot be questioned, even when the evidence claimed in support of Darwin, and for materialism in general, doesn't stack up.

I think, though, that a day will come, in the not too distant future, when it will seem ridiculous that some of us, as late as this, the early 21st century, had to put so much effort into arguing for what, really, should be the obvious. Namely, that Life and Mind on Earth *are intelligent and purposeful phenomena which must, just as obviously, have intelligent and purposeful causes, no matter they be subtle or non-physical.*

For centuries everyone assumed that, one way or another, subtle, spiritual forces gave rise to reality and to us within it all. But, since Darwin, it has become fashionable to believe that we live in a universe without meaning or design. Although it is rarely put quite so bluntly or quite so baldly, we and all living things are, in essence, just lucky coincidences of chemistry.

Yes, a long, and extremely improbable, evolution was needed to get us here, but, at base, that's all we are. This is why it is due to *pure (stupid) chance* that we are now imbued with the extraordinary abilities which we call *sensing and feeling, thinking and knowing, imagination and creativity,* and we are now able, it is claimed, to – ever so cleverly – work out that the mysterious cosmic matrix which gives birth to us consists is, itself completely dumb! It is made of dumb material alone, it is mono-dimensional, not multi-dimensional, and it is *totally without any intelligence or true Life, true Soul or true meaning.*

My "one long argument" in this brief book will be that, although this currently fashionable view of existence appeals to many people as a way to get away from the world's at times oppressive and outdated religions, and to make some sense, at least, of the sufferings on our planet, it *doesn't* work, it can be *shown* not to work, and it's damaging to us and to our science. That it can be demonstrated, quite easily, actually, that living things could *never* arise by mere chemical luck, as Darwin believed might be possible, and that supernatural causation – like it or not – must, somehow, be involved.

Darwin's famous work, *On The Origin of Species,* was one long argument for the allegedly so very extraordinary powers of random causation. By contrast with Darwin's views, you have, here in your hands, a series of clear, powerful, evidence-based arguments for the demonstrable hopelessness of mindless luck to create anything remotely functional, let alone clever human minds, able to explore the world, to discover music, art and mathematics, to be creative and to be able to make at least some sense of reality.

Some may think I argue as I do because I *want* it to be so. No. It's because, probably just like you, I care for what is most likely to be true. I arrived at my conclusions, not due to some dramatic spiritual conversion, but, due to the mounting evidence that Darwin's materialist philosophy did not work. Only years later came my more detailed questionings of his specific evolutionary scheme, including, eventually, the researches for this book and my more recent discoveries as to just how *surprisingly and shockingly shaky* the various evidences, claimed in support of Darwin's theories, really are.

What I can say too is that my arguments for the supernatural and the spiritual as the ultimate causes of Living things, versus Darwin's beliefs in the supreme creative powers of pure chance, do not give weight to any particular religion. None apart from the general claim that the supernatural is real, and that there are various modern forms of repeatable, good quality evidence for it.

Both science and religion make various truth claims but, while religions can demonstrate some of their claims fairly well, for example, that, it's generally better for us to be loving and kind, rather than brutal and cruel, they cannot prove many of their more specific tenets, because they often relate to past events which cannot now be verified. This is where religious *faith* comes in.

Science, though, is not supposed to be a matter of *faith*. It is supposed to *test and to prove* its claims. For example, the claims that all living things are mere accidents, that materialism is true and that it's 'unscientific' to even attempt to inquire into anything *super* natural all need to be tested.

One of the ironies of the current attitude is that, as we'll see, it is really very easy to show that no Living thing could *ever* arise due to some purely random mechanism, with no purpose in mind – because such a random mechanism has no mind. Yet, due to the present *unquestioning faith* in materialism, scientists interested in Life's origins are required to continue to waste their time trying to show how stunningly complex Living things, including our own *bright* Minds, might arise by *pure stupid chance* – despite this being provably impossible.

Why are scientists required to do this? Because to pursue more varied inquiries which might point them in the direction of intelligent causation is, currently, considered to be 'unscientific' and disreputable. No, today, the respectable 'scientific' position is that the mysterious reality which gives birth to us is, itself, completely mindless and unintelligent, and it is considered unscientific to point out the many very obvious problems with this view.

Luckily, the flame to know what is *really* true and how best to live burns *bright* in many of us. It burned *bright* in Darwin and it burns *bright* in Richard

Dawkins. It burns bright in any genuine scientist. But bright and 'right' are not always the same. We can passionately believe whatever we like, that the moon be made of cheese, or that Life could arise by chance, but this, by itself, does not make it true. I'm curious, though, as to how many materialists ever wonder *why* they love the truth or *what* it is, in them, which loves the truth?

What, after all, do the dead carbon and mindless hydrogen atoms of the materialist belief system care about truth? At what point do the atoms of its dry beliefs begin to care and to think, let alone become interested in truth?

Yet Darwin and his companions in the *materialist faith,* for materialism is, in its own grim way, quite like a faith-based religion, ask us to believe, with a very firm *faith,* that, due to wholly *mindless* processes, *non-living* atoms, like carbon and oxygen, accidentally became alive and, eventually, able *to feel and to think* and to discover calculus and to compose concertos, all for no reason or purpose at all. This, they tell us, is the rational view, the scientific view.

Really? No, as I'll to my best to show, this is neither a genuinely rational, nor a genuinely scientific, view. No, it's the materialist view. But is it true?

The following questions will provide our key themes, as we explore contemporary debates about the origins and history of Life through deep time, Darwin's ideas and the true nature of reality.

1. Is Darwin's philosophy – *materialism* – true? Only the physical universe is real, nothing non-physical to play any *causative or Soul in-form-ing* role? Every single spiritual or supernatural experience anyone has ever had is either a hoax or a delusion. Living things may *look* intelligently designed, as many materialists, like Darwin and Dawkins, readily acknowledge, but this is pure *illusion.* No, fundamentally, they arose by *random* processes. Mindless chance + mindless natural laws, themselves random in origin, provide all the apparent creativity. Even complex Living things, Life's DNA CODED software, the brilliant mind of a Plato or Galileo arose by pure, unguided, chance.

2. Or, is materialism *false?* Supernatural phenomena, like *intelligent design,* Life after death, spiritual agencies of various kinds, are real. If so, is there any modern – *non-religious* – evidence for supernatural phenomena which can be investigated by modern, scientific means? Or will scientists always be confined to investigating physical phenomena alone?

3. Many, in science, argue that, even *if* the supernatural is real, a purely materialistic approach, like Darwin's, can, nonetheless, *fully* explain the origin and history of Living things. There is, therefore, no scientific need to make any reference to their supernatural origins, even if real. But is this realistic?

4. Could Life arise by chance? Or is this demonstrably impossible? If materialists, like Darwin, cannot show us, convincingly, *with real evidence* and worked examples, that Life and Mind *really could, and really did,* arise by pure luck, a valid alternative view is the classical intelligent design hypothesis advocated by almost all substantial thinkers prior to Darwin.

5. However Life first started, is there an evolutionary process, as Darwin hypothesized, of one species gradually transforming into another, over and over again, single-cell to worm to fish to amphibian to dinosaur to bird and mammal to ape to human? If there is, could *random genetic mutations* really drive it? Or would such a complex process need to be *intelligently* driven?

6. Some people believe that Darwin was *right* about evolution but wrong as to the *random* means he believed in. Life was evolved, as he described, but by God or Spirit. Are they right? Or, is there is no such process as the kind he argued for, *intelligently guided or random?* What is the evidence of the fossils, embryology, genetics, and today's species? Does it confirm or refute his ideas?

This book's cover also illustrates some of the key ideas we'll consider. Did particles evolve into people by pure chance? Non-living molecules into single cells, single cells into fish, fish into amphibians, dinosaurs into birds, bears into wales, shrews into bats, mammals into people? Are we spiritual beings having physical experiences? Or are we cosmic accidents as Darwin believed?

A brief word here on style. My way of writing may feel a little simple to some, for such a 'serious' subject. If so, it's because I try to approach these matters in a simple but radical way, to question, as a thoughtful child might, various long-held 'truths' of our current science and our wider culture. It took a child, after all, to point out that the gullible emperor, in the famous *Emperor's New Clothes* story, wasn't wearing anything. Perhaps this is also true of Darwin's theory and of the *materialist belief system* which lies behind it.

While I care deeply about these matters, because they relate to some of the most important things we can think about, I have tried not to be too intense towards those on the other side of these arguments. I can understand why, *emotionally,* many adopt Darwin's pessimistic creed. For the same reasons, perhaps, as he did. Life on Earth can be very difficult and we all have to face at least some illness, loss, and, eventually, our departures from this world.

This can make it hard to believe in anything *'more,'* let alone in a good and positive more. Life's deeper sources cannot be seen, at least not physically, and, if there are *afterlife realms* in subtle, non-physical dimensions 'next door,' as most of the world's religious and spiritual traditions maintain, they are not readily discernible to most of us. This makes many adopt Darwin's gloomy creed. But, this, by itself, does not make his ideas true.

Please note, too, that after the first few chapters, where we'll clarify some key terms, words like creation and evolution, natural and naturalism, we'll continue our inquiries in the form of a friendly conversation between two old acquaintances, Alisha and Oliver, on a pleasant country walk.

Firstly, they'll ask, "Could Life ever *arise* by pure chance and is there any evidence that it did?" Secondly, they'll consider Darwin's famous theory and whether Life, however it first arose, could, or did, *evolve* by similar means? Or, they'll ask, have his ideas have become more like a modern *creation myth* for materialism than the good, reliable science they are claimed to be?

Have Darwinism and materialism, which seemed, to many scientists and intellectuals in the 19th and 20th centuries, to promise so much in terms of their potential explanatory power, not only failed to deliver but become science blockers? Are they now leading to *bad science* and to closing down other more interesting lines of inquiry? Has science itself become a kind of pseudo-religion which insists that, "All reality is material or physical. ... Nature is purposeless. Consciousness is nothing but the physical activity of the brain"? [2]

Lastly, in the final chapters, Alisha and Oliver will briefly explore some of the modern – *non-religious* – evidence for intelligent design in Nature and for the supernatural in general. Powerful, modern evidence, some of it repeatable and verifiable, which falsifies Darwin's materialist worldview and shows how science will be so much better off, "freer, more interesting, and more fun," [3] once it has freed itself from these constrictive dogmas.

I realize it can be hard to revisit old beliefs, especially if they were formed long ago, even more so if they fit in with current mainstream views. Why bother to think differently? But, if you have read this far, you are, I'm sure, willing to consider alternative ideas, even if this can be challenging at times.

It's time to get stuck in. I hope you enjoy the read. I have poured countless hours into it, attempting to improve it over many years now, and, if there were any rewards in life for mere effort, it should do very well. But, be its success broad or narrow, I hope these words will reassure at least some of those who, due to Darwin's long shadow, may now doubt that our lives have any meaning or that there is a deeper intelligence to things, especially to living things, than Darwin and modern science have, in recent times, allowed us to imagine.

_____

[2] Sheldrake, R. 2012. *The Science Delusion.* Coronet.

[3] Ibid.

Part I

Demolishing Darwin
Dismantling Materialism

Why all Materialist theories of Evolution,

including Darwin's, fail to explain the

Mysteries of Life and Mind on Earth,

and why, inevitably, they lead to

Bad Science.

1 Disreputable Creation

We have always tried to make sense of our origins and, essentially, there are just two ideas. A *meaningful* creation or creation by pure chance. Either our visible physical universe has subtle *spiritual* levels. Or, it's a cosmic fluke.

If this second view is true, mindless particles of carbon and other elements luckily bumped and clumped together and, after a long and lucky evolution, they became people-shaped, able to *think,* and to ask questions like: *"Who are we? Where do we come from? What is the meaning of our lives?"*

In the west, belief in an intelligent creation comes in two basic forms. The idea of one more or less instant, just a few thousand years ago. This is the biblical literalist idea known as young-earth creationism. Secondly, old-earth creation, where some elements of reality are held to be created more or less instantly, like the Big Bang, and others to be evolved, over deep time, perhaps along Darwinian lines, but with some allowance for intelligence and design.

Belief in creation by pure chance also comes in various flavors. Of these, Darwin's theory is the most famous and it is regarded, by many today, as the materialist school's crowning achievement. Life on Earth is just an amazing cosmic fluke and Darwin proved it. While at one time it was controversial, this idea is now so widespread, in the intellectual west at least, that to express any major doubts about it can be to risk ridicule and to being asked, in a condescending tone of disbelief, "You're not a creationist are you?"

That Darwin's ideas took off in the 19th century was partly because many, in the increasingly scientific west, became less and less willing to be told what to think by the traditional religious authorities. It was also because Bishop Ussher's 17th century idea that the Bible could be read as though it was an early, but accurate, science text – which tells us the precise age of our planet – began to make less and less sense. Modern geological discoveries showed Earth to be far older than the mere few thousand years he had allowed for, based on his very literalistic reading of the timelines in the bible.

At the same time, many became disillusioned with religion altogether and opted for Darwin. He, they felt, had successfully provided a *wholly materialistic* explanation for life, and, for many of his followers, all concepts of the spiritual and the supernatural were truly dead.

While this 19th century reaction against traditional religion, and the supernatural in general, might have been overdone, certain kinds of creationist thinking do contribute to a mainstream atmosphere in which *all* claims about the supernatural and the spiritual can seem suspect. For example, Bishop Ussher's late-medieval, bible-based creationism which is so easy to refute. Earth is clearly much older than the six thousand years he allowed for.

But to go from his medieval idea of creation, that it happened, more or less instantly, just a few thousand years ago, to the belief that there are *no* intelligent or supernatural dimensions to reality at all is a very big jump. This, though, is the dominant position in contemporary science.

The current scientific position is that there is nothing *within or beyond* everyday nature to play any spiritually causative or *Soul* informing role and to claim otherwise, while it may be comforting, is outdated or delusional.

This is the view expressed by well-known science popularizers like Richard Dawkins in books like *The Selfish Gene, The Blind Watchmaker* and *The God Delusion.* The mainstream media also inclines to follow the scientific consensus and the idea tending to hold sway today is that science has,

(a) Shown reality to be *purely material* – with no spiritual dimensions.

(b) The universe started with a massive random Bang *which just was.*

(c) Following this stupendous, random Big Bang, *which just was,* there arose physical laws of such elegant functionality and cosmic constants of such amazing precision, that the best explanation for this – apart from the classical *God or Higher Nature* hypothesis – is multiverse theory.

Multiverse theory describes the notion that perhaps there are numerous universes forming all the time, all caused by some kind of Mindless, Unintelligent and Accidental Universe Generating Mechanism or MUA-UGM, *which Just (mindlessly) Is,* and eventually, by pure luck, such a process *might* give rise to a universe which works and is like the one we happen to be in.

There is no good evidence that this is how existence comes to be but it's a fashionable view. And, currently, there are not many well known scientific dissenters from it. But there are some. Professor John Lennox, an Oxford mathematician, and advocate of intelligent design, is one. He writes,

'Theoretical physicist Paul Davies tells us that, if the ratio of the nuclear strong force to the electromagnetic force had been different by 1 part in 10^{16} [1/10,000,000,000,000,000], no stars could have formed. Again, the ratio of the electromagnetic force-constant to the gravitational force-constant must be equally delicately balanced. Increase it by only one part in 10^{40} [10 with 40 zeros after it] and only small stars can exist; decrease it by the same amount and there will only be large stars. You must have both large and small stars in the universe: the large ones produce elements in their thermonuclear furnaces; and it is only the small ones that burn long enough to sustain a planet with life.' [4]

This is a staggeringly precise condition and to attribute it to mere chance, as many still do, seems odd. But how can those of us who are not mathematicians even begin to grasp just how astonishingly exact this condition is? John Lennox uses an illustration suggested by the astrophysicist Hugh Ross. [5]

'Cover America with coins in a column reaching to the moon (... 236,000 miles away), then do the same for a *billion* other continents of the same size. [Then] Paint one coin red and put it somewhere in one of the billion piles. Blindfold a friend and ask her to pick it out. The odds are about 1 in 10^{40} that she will.'

As if this is not amazing enough, eminent mathematician Sir Roger Penrose calculates that in order to get to the kind of universe we are actually in and one which will actually work,

'the 'Creator's aim' must have been accurate to 1 part in 10 to the power 10^{123}, that is 1 followed by 10^{123} zeros, a 'number which it would be impossible to write out in the usual decimal way, because even if you were able to put a zero on every particle in the universe there would not even be enough particles to do the job.' [6]

Faced with not one, but many such spectacular examples of fine-tuning, it is perhaps not surprising that Paul Davies says, *'The impression of design is overwhelming.'* [7]

So far, however, this kind of intriguing evidence does not much disturb a by now almost unthinking inclination towards chance-based explanations for everything. There is no positive evidence to support the *fashionable view that blind chance can create all,* including our *incredibly* complex bodies and our intelligent minds, but, due to the now very widespread *faith* in pure chance as a supreme creative principle, there is very little discussion as to whether this really works, nor any serious debate as to whether there may be any intelligent, *non-physical* forces which could play any parts in the arisings of our world.

But couldn't there be a spiritual creation which *evolves* rather than one which is instantaneous? Why not? It's an interesting hypothesis and it's the one

[4] Lennox, J. 2009. *God's Undertaker: Has Science Buried God?*. Lion. P. 70.

[5] Ross, H. 1995. *The Creator and the Cosmos*. Navpress. p. 117.

[6] Penrose, R. 1989. *The Emperor's New Mind*. Oxford University Press. p. 344

[7] Penrose, R. 1988. *The Cosmic Blueprint*. Simon & Schuster, p. 203.

held by various esoteric thought schools, like theosophy and anthroposophy. But it's not the mainstream view and it was not Darwin's view.

After all, the very purpose of Darwin's theory was to show that, perhaps, there is no supernatural causation at all. Rather, just random physical events in combination with nature's regular laws could and did lead to all. And, according to this, presently fashionable, view, those laws are, themselves, merely the random byproducts of materialism's MUA-UGM which *'just is.'*

A substitute final source, which is not so different from the classical God, Brahman, Buddha Nature, Allah or Tao of conventional spiritual belief, which, also, *'just is,'* apart from the fact that it contains *no Consciousness, no Will, no Desire, no Creativity, no Intelligence, no Love, nor Purpose of any kind.*

Yet, could such a mindless mechanism *really* generate the kinds of *loving, thinking, creating* Beings we find ourselves to be? After all, isn't it a bit odd to look at another Human being, a playful Dolphin, a soaring Eagle or a beautiful Butterfly, and maintain that all we are observing is a chemical accident? At one time, the answer to this question would have been a universal,*'Yes!'* The classical view being that *Mind came before matter* and the cosmos was a place of *intelligence and spiritual* ideas made visible.[8] A theory of evolution which reflected this understanding, would, unlike Darwin's scheme, include the idea that there may be more to Life than purely physical processes.

If the spiritual is real, there is no reason why an overarching supernatural intelligence, or intelligences, might not evolve its or their creations through various increasingly complex forms. To say that god or spirit can only create instantaneously, not gradually, is not a scientific argument but a *theological* one. If creation is a fact, as evidenced, for example, by the intelligence so widely visible in living things, who says, apart from those who read their scriptures very literally, that *it has to be instant?* There's no law, I know of, to say that *Spirit,* or whatever other labels we may attach to such a force, is not allowed to explore creation *over time* and through various forms of *evolution.*

And for those wary of traditional religious creationism, including bible-based creationism, partly because of its link to other religious teachings which do not sit well with them, to discern that there may be *more* at play in living things, than could be caused by pure chance, does not endorse any particular set of religious or spiritual beliefs. None, at least, beyond the classical idea that there may be more to reality than the numberless chemical coincidences which those who adhere to materialism have, by definition, to believe in.

---

[8] Goswami, A. 1993. *The Self-Aware Universe: How Consciousness Creates the Material World.* Jeremy P Tarcher. Capra, F. 1975. *The Tao of Physics.* Shambhala. Davidson, J. 1992. *Natural Creation or Natural Selection? A Complete New Theory of Evolution.* Element. Theosophical writings of H.P. Blavatsky and Rudolf Steiner among others.

That a meaningful creation may not be *fully* described or understood by any existing religion does not mean that intelligent causation is not possible nor that none of the world's spiritual traditions contains any truth at all. It is ironic that the reasonable 19th century move away from various outdated religious views, like those of Bishop Ussher, seems to have led to a place where our present science rejects *all* ideas of intelligent, non-physical, input.

But to concede that there may, after all, be more to reality in general and to living things in particular, than Darwin and our present science allow us to believe, does not lead to any specific set or sets of religious claims, let alone force us back, as some fear, into some very outdated religious views.

It simply frees science up for a more open minded kind of inquiry than a dogmatic *materialism* allows. Materialism hobbles us and our science. It breezily claims, "Life is nothing special, nothing 'miraculous.' It's a just a 'natural' phenomenon, like wind or rain, and, fundamentally, life arose by pure chance," all without bothering to show us whether this is actually true.

How ironic that we have gone from Bishop Ussher's outdated, medieval biblical-literalist creationism to such widespread and lazy claims about the supreme creative abilities of pure stupid chance which its proponents never bother to substantiate. Absurd, science damaging claims which, as we'll soon see, they *cannot* substantiate.

In *Heretic: One Scientist's Journey from Darwin to Design,* the scientist, Matti Leisola, quotes a colleague, Valtaoja, who, disagreeing with his 'heretical' views, once said to him, very firmly, that *"Life is nothing else than physics and chemistry—mere electricity. There is,"* he says, *"no reason to assume anything supernatural."* [9] This is, currently, the mainstream view, the inquiry limiting view which has lead our present science of origins into a dead-end. This why we will focus on the 'heretical' position, adopted by Leisola and a few other brave scientific souls, that evolution, by the totally mindless means which Darwin, as a materialist, *had* to rely on, could not and *did not* happen.

Of course, the idea that a purely chance based origin and subsequent evolution of life *could* not happen is not the same as showing that it *did* not do so. However, we will argue that, not only is 'life-by-chance' demonstrably impossible, there is no evidence that, despite the impossibility, it *did* occur.

So, if it could not happen and it did not happen, Leisola's 'heretical' premise will prove to be correct. Then, thankfully and at last, our sciences will be freed up for a far more open, freer and more interesting kind of inquiry than Darwin and Valtaoja's restrictive materialist belief system can ever allow.

[9] Lesiola, M. 2018. *Heretic: One Scientist's Journey from Darwin to Design,* Discovery Institute Press.

2 Tricky Evolution

While its meaning might seem obvious, evolution can have some very different, even contradictory meanings. Broadly, it refers to a series of connected changes over time, as opposed to sudden or discontinuous emergences, like the Big Bang or the Cambrian Explosion. And if, according to this definition, we say, "Life has evolved," our statement is hardly controversial. The species *have* changed over time and, many species share features in common – physiological, anatomical, behavioral and instinctual.

The only problem with this definition is that it says very little. It doesn't tell us what drove those changes, nor whether they match Darwin's predictions.

Further, two very different kinds of processes, one random, the other intelligent, can both be called evolution. For example, the evolution of a landscape, while governed by regular laws, is, on the whole, random. But the evolution of our technologies is, serendipitous discoveries aside, intelligence led. It's not just a matter of luck.

Yet evolution is a word now so widely employed that it can become sloppy and misleading. After all, doesn't *everything* evolve? Our technologies evolve and our ideas: vengefulness to forgiveness and the golden rule, monarchy to democracy. Our sciences: earth, wind and fire to the periodic table, flat earth to round earth. So why shouldn't the species evolve? Haven't they changed over time? Isn't that proof of evolution?

No, because *mere change* is not what Darwin meant by evolution. He meant a slow, gradual, rise from the non-existent to the functional, from the simple to the progressively ever more complex, from microbe-to-worm, worm-to-fish, caterpillar-to-butterfly, dinosaur-to-bird, shrew-to-bat, bear-to-whale, ape-to-human, *all due to entirely unguided processes:* pure chance + chemistry.

But our ideas and technologies only evolve due to our *intelligent intentions,* not chance. Landscapes, on the other hand, *do* evolve randomly. And, according to Darwin, like the random evolution of a landscape, the random swirling around of various chemicals + the regular laws of physics and chemistry would, eventually, lead to Plato and Galileo, the brilliant Pythagoras and the astonishing Newton, to Beethoven and Einstein. This belief is, currently, the scientific view, the respectable view. And those who wonder if such a scheme of pure chance, of pure *'unintelligent design,'* can actually work, or whether there is any good evidence for it, are sidelined or ridiculed.

Yet our technologies, all far simpler than any Life form, only evolve as a result of our *thinking* and our ideas working in combination with nature's regular laws. They never arise as the only seemingly clever results of the relevant parts being randomly tossed about, upside down and inside out, while we wait to see what wonderful new gadgets pop out, again and again.

To add to the possible confusion, there are two other kinds of evolution which get referred to in discussions of Darwin's ideas: *micro-evolution* and *macro-evolution.* Confusingly, sometimes misleadingly, they are often conflated. But they describe radically different degrees of biological change.

Micro-evolution is the *uncontroversial* process whereby existing members of a species better adapted to changing conditions, like finches with slightly bigger beaks or peppered moths with darker wings, do better than those members which were better suited to previous conditions. Then, if environmental conditions reverse, those species members with the mix of genes better adapted to the old conditions begin to predominate once more.

There is no evidence, though, that such minor, backwards and forwards changes can lead to the kind of vast *macro-evolutionary* process which Darwin described. This is, however, not generally admitted by Darwin's followers and the critical lines between micro-evolution and macro-evolution are blurred.

Macro-evolution describes Darwin's far more dramatic claim that minor changes *within* a species could, eventually, cause it to change so much as to become an entirely new creature, with new capacities and instincts. A microbe, Darwin felt, might *randomly,* that is without any guiding intelligence behind the process, slowly but very progressively morph into a vastly different worm or trilobite, a fish into an amphibian, a dinosaur into a bird, a bear into a whale, a shrew into a bat, a caterpillar into a butterfly, a speechless ape into a gifted poet or composer of concertos. He felt micro-evolution could lead to – and is, in its essence, no different from – macro-evolution. It's just a matter of degree.

But the fact that finch beaks can vary, within *strict genetic limits,* or that intelligently selecting humans can turn preexisting wolves into many varieties of domestic dog, all of which are still wolves in essence, doesn't prove the modern Darwinian belief that rare, generally harmful, random genetic mutations could turn a single-celled microbe into a *multi-celled* snail or worm, an in-vertebrate worm into a *vertebrate* fish, a water breathing fish into an *air* breathing amphibian, an amphibian into a dinosaur, a scaly, four legged dinosaur into a feathered, two-winged bird or mammals into people.

It is Darwin's concept of *macro* evolution which is being questioned here, firstly, as to the means by which it might take place, random genetic mutations, and, secondly, as to whether there's *any* such process. Changes over time? Yes. But do those changes honestly correlate with Darwin's ideas? We'll see.

3 True Nature of Nature

For Darwin, although the evolution he argued for might *look* intelligent, – and as though design and progress *were* involved, a fact which he readily admitted, – actually, they were not. This was the whole *point* of his theory. It was intended to be a *non-miraculous* explanation of Life's story. All evolution needed were matter + nature's regular laws + irregular chance. Yet, while unguided natural laws can explain the random evolution of a landscape, can they account for Living things, all far more complex than anything we make?

After all, every gadget we create, or the software to drive it, has to be first conceived in a Mind (its idea), for a purpose (its function), its parts created from raw materials and put together in *specific (non random)* sequences to become a working mechanism. For the philosopher, Aristotle, all things, not only man made, could be thought of in terms of these four basic causes:

(1) The idea (Cart, Rock or Life Form): he called this formal causation.

(2) The purpose (what it's for, what it does): he called this final causation.

(3) The materials: material causation.

(4) How it's put together, how it works: efficient or mechanical causation.

And, until modern times, Aristotle's idea, simple as it might seem to us, was respected. Because, although simple, it did, and it does, make sense. For example, the cars in which we drive and the laptops on which we type were all imagined and mined and refined out of the rocks and oils in the ground and put together in *specific and purposeful sequences of parts by intelligent means.*

Each embodies Aristotle's classical four causes: idea, purpose, materials and specific ordering ('efficient causation') to create viable mechanisms. The bodies of bacteria, worms, fish, reptiles, birds, mammals and people are all forged from the raw materials of our planet into thousands of *purposeful (non-random) sequences*, by extremely complex bio-chemical processes. Processes which make our most sophisticated techniques look very basic.

But our current science, fathered in part by Darwin, argues, as did Valtaoja to Leisola, that all things can be explained by reference to *just two* of Aristotle's famous four causes: 'made of' and 'how it works.' They *have* no formal and final causes. A *materialist* stance, sometimes referred to as *naturalism,* where all things are held to have purely 'natural' as opposed to *supernatural* causes – nature is physical only, it's mono-dimensional, not multi-dimensional. For example, "Life is a purely *natural* phenomenon. When conditions are suitable, living things arise *naturally.* They *naturally* evolve, towards ever greater complexity. A progressive evolution is *natural."*

But, although these statements can sound quite pleasing and wholesome, the word *natural,* used in this way, is question begging and it can be lazy. Why? Because it tells us nothing about the *true nature* of Nature. Is she physical alone or does she consist, also, in *Soul and Spirit elements?* Is she intelligent or mindless? The *true* nature of Nature lies at the heart of this debate.

The key to all this is that our science culture has, in recent times, become ever more committed to the idea that it may be possible to explain *all* phenomena, including *Life,* even *Mind,* in purely materialistic terms, purely in terms of the movements of *mindless, non-living,* atomic particles. This is why, today, our sciences and the *materialist belief system* are hard to tell apart.

The materialist conception of nature is very different from the idealist or spiritualist. Materialism's natural world is produced by physical causes alone. Whereas *idealism or spiritualism* sees nature as multi-dimensional and produced, ultimately, *by meaningful, non-physical forces, Life and Soul forces.*

Nowadays, however, our science treats the *super* natural either as non-existent or as irrelevant to its inquiries. Nature is assumed to be wholly physical, and, even if she's *not,* it's felt it won't be necessary to make any references to her *non-physical* components when attempting to make *scientific* sense of our world. But does this genuinely work and is it genuinely scientific?

Would it not be like desert islanders saying of a previously unknown machine, like a Car, suddenly appearing on their island: "Regardless of whether this mechanism has intelligent sources, we believe we can *fully* explain it in purely natural terms? Under the right conditions, it could have 'fallen together' by pure chance. However, even if intelligence *was* involved in its arising, we will still be able to explain it *as if* no intelligence was involved."

But, just as this won't work for the Car, it won't work for Living things either, if there is more to them than (dumb) matter + (dumb) luck. Equally, if we have *immaterial Minds and Souls* which continue, after death, into spiritual dimensions beyond the physical, are we not, still, part of an *overarching Nature* which is multi-dimensional and contains non-physical elements?

This is why it's more straightforward to use the terms *materialist and materialism* rather than *naturalist and naturalism* to describe the way of thinking which is, currently, considered 'scientific.' Materialism is clear. It describes the popular belief that only the physical is real. Naturalism is less clear, because some consider nature to be purely physical, while others see her as multidimensional and including *Subjective, Soul and Spirit* elements.

4 Is Life Just an Amazing Cosmic Accident?

From here on, dear reader, we'll continue our inquiries by listening to two friends, Alisha and Oliver, as they walk and talk on a pleasant day out in the country. We join them as Oliver, the narrator, sums up the discussion so far.

'So, according to our current science, Ali, especially since Darwin, Life is thought to be just a very lucky chemical coincidence. It is no longer considered to be a *miracle* in the true sense, to be the meaningful product of meaningful causes, of a *supernatural* intelligence which, *mysteriously, just Is.*

'That's sad.' Said Ali. 'It also implies that there's no Life after this Life.'

'Yes, but what if it's true?'

'You mean we'd just have to resign ourselves to being mere cosmic accidents, without purpose or meaning, and, when the chemicals stop that's it?'

'Yes. But, as would-be scientists and truth seekers we must care for what's *true,* not merely what we hope to be. On the other hand, if this idea is false, and it can be shown to be, Darwin's materialist paradigm will be refuted, our sciences will be freed up for fresh inquiries into origins, and life will become so much broader, more interesting to explore, and more meaningful again.

'But are there any ways to demonstrate that Darwin might be wrong?'

'Yes, there are, plenty actually. But, firstly, the idea that Life is just an accident which evolves by accident, was not, despite the misleading title of his famous book, *On the Origin of Species,* a formal part of Darwin's theory.'

'Why was the title misleading?'

'Because Darwin did not truly explain the origin of species at all. He simply *assumed* that Life might have started in the form of an original microbe which *might* have varied until it *might* have become all today's species. But, given his inclinations to materialism, it's no surprise to find that he wrote to a friend:

> 'But if (and oh what a big if) we could conceive in some warm little pond with all sorts of ammonia and phosphoric salts – light, heat, electricity etc present, that a protein compound was chemically formed, ready to undergo still more complex changes'[10] [and, Darwin imagined, accidentally become a tiny Living Cell, 'the most complex system known to man.'[11]]

---

[10] Darwin to J.D. Hooker. 2.2.1871.

[11] Davies, P. 1998. *The Origin of Life.* Penguin.

'And what, back in Darwin's day, was a daring, even disreputable speculation to utter, has become, by now, more or less a scientific dogma.

'That Life arose by chance, as he suggested in his letter to his friend?'

'Yes. Many people now believe that life really could have started that way. But, for a hypothesis to be scientific, Ali, it must be capable of testing, otherwise it's just speculation. So, let's test the by now very popular claim:

"Life arose by pure chemical luck. Darwin showed that if we allow for enough lucky rolls of the chemical dice, anything can be created."

'Let's see if this is true. In Darwin's day, it was thought that the kind of life which might play the role of his hypothetical universal common ancestor microbe was relatively simple. This is why he and some of his colleagues wondered whether Living Cells might arise spontaneously. Darwin's famous contemporary, Ernst Haeckel, described the Living Cell as a 'homogeneous globule of protoplasm.' He thought cells were pretty simple things, so he believed they *might* arise by pure chance. He was, however, spectacularly wrong.'

'In what way was he wrong?' Queried Alisha.

'Because Cells, he did not realize, are *unbelievably* complex! Did you know that, although smaller than sand grains, each tiny cell is vastly more complex, at its individual level, than the much larger organs, like lungs, stomachs and kidneys, which it helps to build? Every cell is, *literally,* like a gigantic, micro-miniaturized, super-computer-controlled chemical factory, far, far more complex than *any* of our factories or even whole groups of them!'

'But Darwin believed they might have arisen by chance?'

'Yes, because he had *no idea* just how complex cells were.'

'Is there any evidence to support this idea?'

'No, none, absolutely none. And there is no evidence that mere chemical coincidences could give rise to two of the basic features which distinguish all living things and, indeed, all the mechanisms we make.'

'Namely?'

'All Life Forms are both *complex and specific.'*

'Go on...'

'Well, not only are living things very complicated but *that complexity is not random.* Imagine a Car or Computer. Its parts have to be assembled in a very *specific and purposeful* order or it won't work. Mere chance cannot give rise to this kind of non-random complexity. It is demonstrably impossible.

'This is why, as everyone knows, if we randomly tumbled Car, TV or Computer parts around – and around – for as long as you like, we'd never get a working machine. Yes, the mix would be extremely complex, but it would be random. It would contain no functionally arranged sequences of parts.

'Any natural ordering which might arise from the random tumbling would, at best, cause the heaviest parts to sink to the bottom of the mix. Regular physical laws alone do not have the power to generate the highly complex, irregular, yet totally specific sequences of software and parts which characterize all computer codes, all mechanical devices and all Living things.

'Random swirling about of parts will never give rise to a specifically ordered mechanism, nor to the **DIGITALLY CODED** software to drive it. In what we can think of as regular, everyday nature it is well established that chemical accidents can only give rise to *random* complexity, as in a forest fire or avalanche, or, at times, to specific, but relatively simple patterns, as in crystal formation. Regular, repetitive patterns which can be described by simple algorithms along the lines of **ABC + ABC + ABC + ABC**.

'As origins of life researcher, Lesley Orgel has explained,

> *'Living things* are distinguished by their *specified complexity.* Crystals such as granite [although specific, consist in relatively simple patterns so,] fail to qualify as living because they *lack complexity;* mixtures of random polymers [although highly complex] fail to qualify because they *lack specificity.'* [12]

'But a regular physical law which can generate simple, specific sequences, like **ABC + ABC + ABC**, cannot generate the far more complex sequences required by instructions like **"TURN THE SWITCH ON HERE."** A complex, irregular, but non-random and specific sequence like this is *physically indeterminate.* Due to its meaningful content, it cannot be determined by physical laws alone. It can only be produced by intelligence.

'Similarly, a DNA CODING sentence like **TTAAGGCC + ATAGGAGC CAACGGTT + TTAAGGCC + CCTTGTAC** cannot be produced by regular physical laws alone, because, not only is it complex, it is irregular. Not only is it irregular and specific, it is purposeful. Such a sentence, it can be mathematically demonstrated, cannot be produced by chance.

'Further, as biochemist Michael Behe points out,

> 'There is *no* publication in the scientific literature in prestigious journals, specialty journals, or books that describes how molecular evolution of any real, complex, biochemical system either did occur or even might have occurred. *There are [many] assertions that such evolution occurred, but absolutely none are supported by pertinent experiments or calculations.'* [13]

---

[12] Orgel, L. 1973. *The Origins of Life*. John Wiley; p. 189. Emphasis added. I was guided to Orgel by Stephen Meyer in his important book on intelligent design, *Signature in the Cell*.

[13] Behe, M. 1996. *Darwin's Black Box*. The Free Press, p. 185. Emphasis and text in square brackets added.

'Ok, but what about the spark and soup experiments of the 1950s?' Asked Ali. 'Isn't Miller and Urey's famous 1953 experiment often cited as evidence that life could have started by purely random means? By pure chemical luck?'

'Yes, but it's a misleading myth. Remember, all the experiment generated was a few of life's basic building block amino acids. But a few random bricks don't make a building, let alone a complex one. Secondly, it's now believed Miller and Urey's assumptions about the earth's early atmosphere were flawed and had they correctly simulated it, by including – rather than excluding – oxygen, those few amino acids would not have been produced at all.

'In any case, building blocks, unguided by higher level forces, will never lead to anything but random piles of blocks or, at best, relatively simple patterns. But simple patterns cannot give rise to the many, not just a few, specifically ordered sequences of parts of even the simplest of living mechanisms. Note, too, that as molecular biologist Michael Denton writes:

> 'Instead of revealing a multitude of transitional forms through which the evolution of the cell might have occurred, [modern] molecular biology has served only to emphasize the *enormity* of the gap. We now know not only of the existence of a break between the living and nonliving world, but also that it represents the *most dramatic* and fundamental of all the discontinuities of nature. Between a *Living Cell* and the most highly ordered nonbiological system, such as a crystal or snowflake, there is a *c h a s m* as vast and absolute as it is possible to conceive.' [14]

'In fact, as biochemist Michael Behe points out, even a simple mousetrap with its specifically and purposefully ordered parts could never fall together by chance. It can only be assembled by intelligent agency. If we are generous, Darwin, and his steam era contemporaries, can be forgiven for imagining that Life might arise by 'natural chance.' We, though, who know so much more than they, have no excuse for continuing to think as they did. They had slim knowledge of modern probability theory and the severe limits it imposes on the creative capabilities of random causation.

'They, also, had *no idea* just how breathtakingly complex DNA CODED Life would eventually turn out to be. They had no idea that,

> 'Although the tiniest bacterial cells are incredibly small . . . each is in effect a veritable micro-miniaturized factory containing thousands of exquisitely designed pieces of intricate molecular machinery, made up altogether of one

---

[14] Denton, M. 1985. *Evolution a Theory in Crisis.* Burnet Books, pp. 249–250, emphasis and text in square brackets added.

hundred thousand million atoms, *far more complicated than any machinery built by man* and absolutely without parallel in the non-living world.' [15]

'They had no idea that,

'To grasp the reality of life as it has been revealed by molecular biology, we must magnify a cell 1000 million times until it is 20 km in diameter and represents a giant airship large enough to cover a great city like London or New York . . . On the surface of the cell we would see millions of openings, like the portholes of a vast spaceship, opening and closing to allow a continual stream of materials to flow in and out. If we were to enter one of these openings we would find ourselves in a world of ***supreme technology and bewildering complexity***. We would see endless highly organized corridors and conduits branching in every direction away from the perimeter of the cell, some leading to the ***central memory bank in the nucleus and others to assembly plants and processing units***. The nucleus itself would be a vast spherical chamber more than a kilometer in diameter, resembling a geodesic dome inside of which we would see, all neatly stacked together in ordered arrays, the miles of coiled chains of the DNA molecules. A huge range of products and raw materials would shuttle along all the manifold conduits in a highly ordered fashion to and from all the various assembly plants in the outer regions of the cell.

'We would wonder at the level of control implicit in the movement of so many objects down so many seemingly endless conduits, all in perfect unison. ***We would see all around us, in every direction we looked, all sorts of robot-like machines*** ... We would see that nearly every feature of our own advanced machines had its analogue in the cell: ***artificial languages and their decoding systems, memory banks for information storage and retrieval,*** elegant control systems regulating the automated assembly of components, ***error fail-safe and proof-reading devices used for quality control,*** assembly processes involving the principle of prefabrication and modular construction . . . However, it would be a factory which would have one capacity not equaled in any of our own most advanced machines, for it would be capable of replicating its entire structure within a matter of a few hours ...' [16]

'This kind of complexity, Ali, is not good evidence for "unintelligent design" for the mere ***chemical luck*** of the kind that so many now believe in. It

---

[15] Denton, M. 1985. *Evolution a Theory in Crisis.* Burnet Books, p.250, emphasis added.

[16] Ibid. p 328, emphasis added.

is intellectually lazy, to assume, without evidence, as so many now do, that this incredible intracellular organization must have, 'somehow or other,' just evolved, by pure chance, based on nothing more than the physical invisibility of life's more subtle causes and the fact that Darwin noticed that clever people, with clear, pre-defined search criteria in mind, could cause already incredibly complex pigeons or cattle to vary to some, *limited, species-bound,* extent.

'But to extrapolate from such observations of uncontroversial, artificial micro-evolution, that incredibly complex living things might arise by chance in the first place, then evolve from there, also by chance, was totally unjustified by the data, then or now.

'It does seem odd to put so much faith in pure chance,' said Ali.

'Yes, but it flows, Alisha, from a rejection, in the modern west, of all ideas of the supernatural. Why? Because people began to realize that the world's traditional creation stories could not be treated as science and because life's deeper sources cannot be seen physically; they can only be inferred, using logic, or known, to some extent, by extrasensory perception.

'But to go, as Darwin did, from, "The Bible is not a literal description of natural history," "Yes, we know that, it's a religious book. It's not science." To, "So all ideas of the spiritual and the supernatural are false and life must just be nothing more than a cosmic accident!," is a very big and unnecessary jump.

'To ascribe the stunning complexity of living things to 'unintelligent design,' to mere chance because we now realize that the world's traditional creation stories are not 'science' is unreasonable. In any case, facts trump theory. The facts do not support Darwin's key idea that pure chemical luck can create anything functionally clever. Can you think of any cases of mere 'chemical luck,' of mere 'natural accident,' creating anything remotely intelligent, let alone a working mechanism of some sort?'

Alisha reflected, 'No, natural accidents make me think of rusts, frosts, storms, avalanches, forest fires and other random events of various kinds.'

'Yet mainstream western thought, Ali, which so prides itself on its rationality and its commitments to evidence based science, is in peculiar thrall to this deeply unempirical idea. The famous astronomer, Carl Sagan, said, 'extraordinary claims require extraordinary evidence.' Fair enough, I agree.

'The trouble is, there is *no evidence at all,* let alone any extraordinary evidence, to back up our current science's claims for the supposedly so very amazing creative powers of chance causation. An idea made popular, today, partly because some people began to put about the following urban legend, with *no* rational basis and *no* substantiating evidence of any kind, that,

'Oh, just give it enough time and even a roomful of monkeys, typing totally randomly, would, eventually, produce the complete works of

Shakespeare! Such is the power of random causation! It's just a question of time and probability. Pure luck can build anything, even digital CODES or robots programed with Artificial Intelligence or Living Cells of stunning complexity, even entire Plants, Animals and Humans, orders of magnitude more complex than any software, computer or robot we make.'

'It's funny, Ali, but many believe this stuff. Yet it's nonsense statistically.'
'Why do people take it seriously then?' Mused Alisha.

'Firstly, because of a steep decline in the west of belief in the miraculous and, secondly, because scientists, today, are *trained* to assume that wholly unintelligent processes can give rise to everything, no matter how complex. But the creative capabilities of pure chance have been grossly overestimated. It is easy to demonstrate that mere luck cannot create functional complexity.

'In his book, *Genesis and the Big Bang,* physicist Gerald Schroeder looked at the odds of monkeys randomly typing out *just one* Shakespeare poem, the beautiful, 'Shall I Compare You to a Summer's Day?' He wrote,

> 'There are 488 letters in the sonnet . . . The chance of randomly typing the 488 letters to produce this one sonnet is one in 26 to the 488th power, or one in 10 to the 690th power. The number 10 to 690 is a one followed by 690 zero's! . . . [Yet] since the Big Bang, 15 billion years ago, there have been only 10 to the 18th power number of seconds . . .
>
> [In fact, at] one random try per second, with even a simple sentence having only 16 letters, it would take 2 million billion years (the universe has existed for about 15 billion years) to exhaust all possible combinations.' [17]

'An article in wikipedia, *Infinite Monkey Theorem,* makes the same point,

> 'Even if *every proton* in the observable universe were a monkey with a typewriter, typing from the Big Bang until the end of the universe ..., they would still need a ridiculously longer time - more than three hundred and sixty thousand *orders of magnitude* longer – to have even a 1 in 10^{500} chance of success[fully typing out *Hamlet].* . . .
>
> 'In fact there is **less than a one in a trillion chance of success** *that such a universe made [entirely] of monkeys could type any particular document a* **mere 79 characters long.'** [18]

'When it comes to living things, leading modern intelligent design proponent, William Dembski, shows how life, also, could *not* arise by chance.

[17] Shroeder, G. L. 1996. *Genesis and the Big Bang.* Bantam Doubleday Dell

[18] Wikipedia

He explains that even if the *entire* universe had consisted in a pre-life chemical soup and even if it had been *deliberately* set to generate random amino acid sequences by using every last particle, and even if it had been doing nothing else, since the Big Bang, it would never lead, by pure luck, to *just one* of the thousands of specific protein sequences needed to make cellular Life possible.'

'Why not?'

'Because the laws of probability forbid it. And, even if one of these sequences ever did arise – by chance – with no cellular membrane to protect it, it would be as rapidly destroyed by all the other chemical-reactions going on around it, 24/7. Remember, it's the DNA *inside* the Cell which codes for the very membrane which protects it from the *outside*. There's no way the DNA inside could arise, by chance, to code for the membrane which protects it from the outside. These factors are irreducibly complex. They can only arise together. It is demonstrable that they could never arise by 'chance.'

'Nonetheless, Dembski explains that a mechanism's non-random patterning, its 'specificity,' should be high before mere luck can be completely ruled out as a possible cause. He explains, however, that, at a certain point, a particular sequence of information or of parts, made, in the case of cells, of polypeptide chains, *"becomes too long, too improbable, too complex to reasonably attribute to unspecified, unintelligent chance."* [19]

'In their book, *Intelligent Design Uncensored,* Dembski and Witt argue that scientists have the skills, today, to scientifically detect evidences for *intelligent* design versus unintelligent (accidental) design. So, they ask, is the arrangement of the parts of the most basic cell complex enough to pass a rigorous mathematical test for detecting intelligent specification (deliberate design) in nature versus accidental [unintelligent] design? They write,

> 'How improbable does a specified thing [a specific sequence of coding symbols or parts] have to be before we can know it was designed? One chance in ten is obviously too low a threshold. One chance in a hundred is too low. How high is high enough? For our purposes, we don't mind if the test occasionally misses design. We just want to make certain the test doesn't say something was designed that wasn't. This means we want to set the bar very high, meaning **the thing in question will have to be extremely improbable to pass our design test.**' [20]

[19] Dembski, W and J. Witt. 2010. *Intelligent Design Uncensored.* IVP, p. 66.

[20] Ibid. p.67, text in square brackets and emphasis added.

'This is why, in his book, *The Design Inference* [21] Dembski sets the threshold, for design, extremely high, at 1 in 10^{150}, 1 followed by 150 zeros.

'How do they get to that vast number?' Asked Ali, intrigued.

'Well, in the known universe there are about 10^{80} elementary particles and Planck Time shows that matter can change from one state to another no faster than 10^{45} per second or 10,000,000,000,000,000,000,000,000,000,000,000,000,000,000,000 per second. If we now suppose, they write, 'that any event in the universe requires the transition of at least one elementary particle . . . then the total number of events throughout cosmic history [14 billion years] could not have exceeded 10^{80} x 10^{45} x 10^{25} => 10^{150} [10 followed by 150 zeros].'

'So, Dembski and Witt ask: "Is the arrangement of the parts of the most 'basic' Cell – which is highly specific," and is, in essence, an extremely complicated, "self-reproducing factory for producing functioning hardware and software.." [22] – complex enough to pass this extremely rigorous test for detecting intentional design in Nature versus mere chance? They write,

> 'Scientists calculate that a cell with just enough parts to function in even a crude way would contain at least 250 genes and their corresponding proteins. The odds of the early Earth's chemical soup randomly burping up such a micro-miniaturized factory are unimaginably longer than 1 chance in 10^{150}, [a probability boundary beyond which to attribute anything to random causation becomes completely unreasonable because it's impossible].' [23]

'Dembski and Witt show that, just as with *the irregular but specific and purposeful sequences* in Shakespeare's poetry, or the irregular but specific sequences in which a computer, car or TV's parts are put together, living things could not possibly arise by mere luck. It's not just highly improbable, as all agree, it's impossible, as many, so far, still struggle to see.

'So why, Ollie, is the idea that mere chance can create all still so popular, especially in the scientific community, despite it being so intellectually faulty?'

'Partly because many scientists, today, find the alternatives unpalatable.'

'Why though?'

'Well, for one, because their concepts of the supernatural and the spiritual, which they reject on the grounds of unbelievability, tend to be very simplistic, you know, the medieval, Father Christmas type God which no one should believe in and no one needs to believe in.' We walked on.

---

[21] Dembski, W. 2006. *The Design Inference: Eliminating Chance through Small Probabilities. Cambridge Studies in Probability, Induction and Decision Theory.* Cambridge University Press.

[22] Ibid., p. 71.

[23] Ibid. p. 71. Text in square brackets added.

5 Fabulous Chance or Fat Chance?

Enjoying the freshness of the morning air, we continued our conversation. 'Although materialism is a belief system, Ali, not proven fact, it gets a lot of its support today from the scientific community. The influential community which, alongside Darwin, maintains that Life is probably a purely physical phenomenon. There's nothing supernatural or genuinely *miraculous* about it.

'Yes, many in science agree that the world's religions can continue to bring us useful moral codes and comfort to millions, but, when it comes to Life's *true* origins, they consider the religious and the spiritually minded mistaken in perceiving nature to be 'intelligently designed' or 'purposeful' in any way.

'For example, the famous materialist, Richard Dawkins, writes in his book, *The Blind Watchmaker,* "[Biology is] the study of complicated things that give the *appearance* of having been designed for a purpose" but, actually, he believes, "the evidence of evolution reveals a universe without design."

'For Richard Dawkins, like his 19th century hero, Charles Darwin, all living things, even our very minds, can be explained by wholly unguided natural processes. This is why Dawkins feels that "Darwin made it possible to be an intellectually fulfilled atheist." [24] Darwin had shown, he believes, that materialism and atheism were *scientifically* justified. There was no longer any need to appeal to the supernatural to explain Life and Mind on Earth.

'Yet, intriguingly, in a letter to the distinguished astronomer John Herschel, Darwin himself wrote,

> 'One cannot look at this Universe with all [its] living productions & man without believing that **all has been intelligently designed;** yet when I look to each individual organism, I can see no evidence of this.' [25]

'That's interesting, Ollie. Darwin, it seems, unlike his most famous modern fan, Richard Dawkins, was open to the classical idea that "all has been intelligently designed," but then reversed himself?'

'Well, he was conflicted. On the one hand, like so many before him, and so many since, he could see evidences for intelligence and design all around. On the other hand, it is not possible to see nature's deeper causes physically and

---

[24] Article by J. Wells. https://evolutionnews.org/2008/04/is_the_science_of_richard_dawk/

[25] Letter to J.F.W. Herschel, 23.5.1861. www.darwinproject.co.uk

life can be challenging and difficult to make sense of, so, on balance, like his most famous modern admirer, he inclined to materialism.

'Remember, Darwin's arguments were as much *philosophical* and emotional as scientific. If you incline to a disbelief in anything beyond the physical, you *have* to come up with entirely materialistic theories in your attempts to explain existence. You have no other option. Interestingly, though, a while later, another famous astronomer, Sir Frederick Hoyle, wrote:

> **'Life cannot have had a random beginning** . . . The trouble is that there are about two thousand enzymes, and the chance of obtaining them all in a random trial is only one part in $10^{40,000}$ [1 followed by 40,000 zeros] an outrageously small probability that could not be faced even if the whole universe consisted of organic soup.[26]

'He added,

> 'Once we see, however, that the probability of life originating at random is so utterly minuscule as to make it *absurd,* it becomes sensible to think that the favorable properties of physics on which life depends are in every respect deliberate ... . It is therefore almost inevitable that *our own measure of intelligence must reflect ... higher intelligences ... even to the limit of God* ... such a theory is so obvious that one wonders why it is not widely accepted as being self-evident. *The reasons are psychological rather than scientific.* [27]

'In fact, Eastman and Missler write,

> 'Hoyle's calculations may seem impressive, but *they don't even begin to approximate the difficulty of the task*. He only calculated the probability of the spontaneous generation of the proteins in the cell. He did not calculate the chance formation of the DNA, RNA, nor the cell wall that holds the contents of the cell together.

> 'A more realistic estimate for spontaneous generation has been made by Harold Morowitz, a Yale University physicist. Morowitz imagined a broth of living bacteria that were super-heated so that all the complex chemicals were broken down into their basic building blocks. After cooling the mixture, he concluded that the odds of a single bacterium re-assembling by chance is one in 10 to 100,000,000,000. This number is so large that it would require several thousand blank books just to write it out. To put this number into

---

[26] Hoyle, F and N.C. 1981. Wickramasinghe, *Evolution from Space*. J.M. Dent & Sons.

[27] Ibid. pp. 141, 144, 130

perspective, *it is more likely that you and your entire extended family would win the state lottery every week for a million years than for a bacterium to form by chance!* [28]

'Presumably, then,' said Alisha, 'those, like Dawkins, who continue to assert that wholly mindless processes could create everything, including their own *bright minds!,* do so for philosophical reasons, not ones of science?'

'Yes, as Frederick Hoyle noted. But is it not hard to take seriously, Ali, an intellectual position which, in the name of rational, logical, science, claims that when we are looking at a soaring Bird, a swooping Bat, a bustling Beetle or a gently flapping Butterfly, what we are *really* observing is, in essence, a meaningless coincidence of chemistry moving through the air?'

'It does sound pretty unreasonable if you put it like that.'

'But it is, literally, what these beautiful things would be if the materialist worldview were true. Not many people realize this, Ali, but it is a basic tenet of our current science that there is *no intelligence or teleology (purpose)* in Nature.

'According to this fashionable, if peculiar, view, your skillful hands and your *bright* mind are just lucky accidents of a wholly mindless and 'purposeless' evolution. The same for our blood and our muscles, for our brains and our hearts, for Life's complex **DNA CODING** system and the thousands of other mechanisms which make life possible.'

'That doesn't make sense.' Exclaimed Alisha. 'Living things are so clearly full of purpose and they are packed with incredibly clever functionality.'

'Yes, but the current mainstream scientific position is that Life is just an accident which evolved by accident. Most scientists presently working in natural history are, literally, taught not to allow that anything *meaningful or supernatural or intelligent* may be giving rise to Living things.

'Our present science finds it hard to imagine that supernatural intelligence, however we may label it or think of it, might wish to explore and to express creation in many different ways, not just through our senses but through the senses of many other organisms, all with their own qualities of awareness.

'Did you know that there are dogs who like skate-boarding, ducks who like feeding fish, and parrots who like dancing, in time, to music? They can be seen on YouTube. Who are we to say they have no *Soul or Spirit,* no inner reason for existing and playing, just because we don't know what they're 'for'?

'No, the current view is that they are, in essence, mere coincidences of chemistry. Our present science, as Valtaoja put it so clearly, takes the view that there is *no* intelligence or purpose to anything in nature. It's all accidental.

---

[28] Eastman, M & C. Missler. 1995. *The Creator Beyond Time and Space.* Emphasis added.

'Darwin, though, didn't know that, as biologist Michael Denton explains, for us to build a scale model of just one cell, at one tennis ball sized atom per second would take us, literally, 100s of thousands of years. Yet each cell, of which there are trillions in our bodies, achieves this at supercomputer speeds in just a few hours. This, Ali, is not good evidence for the workings of pure chance or 'unintelligent design.' Yet, right now, this is the mainstream position.

'In his foreword to the 20th anniversary edition of Philip Johnson's *Darwin On Trial,* professor of biochemistry Michael Behe notes that,

> 'the situation for [wholly] materialistic origin-of-life theories has gotten substantially worse [since Johnson's book was first published in 1991]. Broadly speaking, for decades there have been two categories of [materialist] origin-of-life theories: the "metabolism-first" view, where metabolic reactions in an enclosed space precede the occurrence of genetic material; and the "genetics-first" view, where a DNA-like polymer that is capable of carrying information precedes cells. The partisans of both camps have offered ***devastating criticisms*** of each others views, so that none are left standing... the only reason at present to believe in a [wholly] materialistic origin of life is if one holds it as a postulate that life ***must*** have had a materialistic origin.' [29]

'Such a postulate, though, is question begging. What if life did not have a purely materialistic origin? What if life really *is* miraculous – in the true sense?

'However, as scientists are currently trained to ignore this possibility, some have argued that the INFORMATION RICH molecules upon which life is based might arise where a non-living mixture of relatively simple proteins and RNA molecules reached a point at which the mixture might start to self-replicate. This hypothesis is called RNA World origin-of-life thinking.'

'But why would such a stage arise?' Mused Alisha. 'In any case, mere replication doesn't mean something is alive, nor does it explain what Life truly is... does it? Crystals can replicate but they are not alive in the way we are.

'I agree, Ali. RNA molecules already contain COMPLEX CODED INSTRUCTIONS, the origins of which remain unknown. As William Dembski explains in *No Free Lunch: Why Specified Complexity Cannot Be Purchased Without Intelligence,* there is no such thing as *'free'* software or *'free'* CODE.

'This is a problem which no purely materialistic theories of Life can overcome. Codes always come, they *have* to come, from the intelligent, not the mindless. Yet all Life forms are run by digital DNA CODES which make even our most sophisticated IT look basic. Despite this, the current scientific

---

[29] Johnson, P. 2010. *Darwin on Trial.* IVP Books, p. 16, emphasis and text in square brackets added

view is that the disorganizing forces of randomness, chaos and entropy would not be a problem for a purely chance-based origin for Life.'

'Why not?'

'Because, it is held, the energy of the Sun would provide Life with all the 'free fuel' needed to drive Darwin's evolutionary process forward.'

'In other words,' said Alisha, 'all Life needed to get started, and to evolve from there, were free laws + free energy + free chemicals + lots of random mixing, but no need for any *intelligence or specific ordering* at any stage?'

'Correct. But this is like imagining that monkeys could randomly type up some 'free' Shakespeare, that a chemical reaction like a random forest fire, with plenty of 'free' fuel, could assemble anything other than burnt wood or a tornado blowing through a junk yard could assemble a 'free' jumbo jet.

'Living cells, which are like tiny chemical factories, of a truly astonishing complexity, depend on incredibly sophisticated DNA CODED INSTRUCTIONS to drive them. Modern probability theory shows us that these kinds of instructions *cannot* arise by chance. It's not merely highly improbable, as all agree, it's *impossible,* as many, oddly, still struggle to see.

'As Dembski and Witt point out, the chances of an imaginary pre-Life soup accidentally generating the protective cell wall for the contents of the Cell at the same time as the DNA needed to CODE for that same cell wall are far, far beyond improbable. It results in a number so high that: "it stretches the ability of biochemists to calculate it. The best they can do, is set a lowest possible figure, which surpasses by untold trillions of trillions of trillions of times the universal probability bound of 1 chance in 10 to 150." [30]

'Even Jacques Monod, who was himself a committed materialist and originator of the famous phrase that solely 'chance' (chemical coincidences) + 'necessity' (regular physical laws) can explain everything in Nature, wrote:

> 'we have no idea what the structure of a [hypothetical, accidental] primitive cell might have been. ... the simplest cells available to us for study have nothing "primitive" about them. ... The development of the metabolic system, which ... must [by chance] have "learned" to mobilize chemical potential and to synthesize the cellular components, poses Herculean problems. So also does the emergence of the selectively permeable membrane without which there can be no viable cell. But the major problem is the origin of the genetic CODE and its translation mechanism. ...

> 'The code is meaningless unless translated. The modern cell's translating machinery consists of at least fifty macromolecular components WHICH

---

[30] Dembski, W and J. Witt. 2010. *Intelligent Design Uncensored.* IVP, p. 72.

ARE THEMSELVES CODED IN DNA: THE CODE CANNOT BE TRANSLATED OTHERWISE THAN BY PRODUCTS OF TRANSLATION [emphasis in original]. It is the modern expression of omne vivum ex ovo [Life only comes from Life, so which came first, chicken or egg?].'[31]

'In *The Naked Emperor: Darwinism Exposed,* Dr. Antony Latham makes a similar point. He writes,

'Far too often I hear experts confidently postulate about the first cell membranes. They know that the first life had to have a protective coat [an outer wall] to hold the DNA and all the incredibly sophisticated molecular machinery inside the bacteria. ... [yet] the actual details of the membrane must be coded for in the DNA of the cell.' [32]

'As both the materialist, Jacques Monod, and modern intelligent design thinker, Antony Latham, point out, how could the Cell's complex DNA coded software arise – by unguided processes – without first having the protective cell membrane around them, which *it* codes for? The cell's DNA and its outer wall are *irreducibly complex*. They have to arise together or not at all. John Lennox, a professor of mathematics at Oxford, writes:

'The evidence, as we've seen, is that: 1. Life involves a complex DNA database of digital information. 2. The only source we know of such language-like complexity is intelligence. 3. Theoretical computer science indicates that unguided chance [coincidences] and necessity [regular laws] are incapable of producing semiotic (language-like) complexity.'

'Intelligent design advocate Werner Gitt comments that: "It has never been shown that a coding system and semantic [meaningful] information could originate by itself . . . The information theorems predict that *this will never be possible. A purely material origin of life is thus [ruled out]"* [33] In view of this, John Lennox writes:

'So, on the basis of making a scientific inference to the best explanation one would have thought that scientists would prefer an explanation that explains a given phenomenon over an explanation that does not. The fact that this is not the case in thinking about the origins of

---

[31] Monod, J. 1972. *Chance and Necessity.* Collins. pp 134–135, except where indicated, emphasis and text in square brackets added.

[32] Latham, A. *The Naked Emperor: Darwinism Exposed.* London: Janus, p. 13.

[33] Gitt, W. 2006. *In the Beginning Was Information: A Scientist Explains the Incredible Design in Nature.* Master Books, p. 124; emphasis and text in square brackets added.

life shows that an a priori **materialism can produce a profoundly anti-scientific attitude** – [an] unwillingness to follow evidence where it clearly leads simply because one does not like the implications of so doing.' [34]

'So science is now being held back by philosophical assumptions which hold materialism to be true even if it may not be?' Asked Alisha.

'Yes, but, don't worry, Ali, we will eventually win this vital debate.'

'How? It seems science is now so committed to the materialist worldview.'

'Because, Ali, gradually, more people are starting to realize that the mere stupid dumb luck, which materialist thinking so fervently asks us to believe in, simply doesn't have the astonishing creative powers which it so futilely attempts to attribute to it. This means it's not unreasonable to revert to the classical view that there is *intelligence* behind living things.'

'Ok, Ollie. But doesn't that have religious implications? Don't many people today feel that science stands in opposition to all things supernatural or spiritual? They see science and the miraculous as mutually exclusive.'

'Yes, many do see it that way. But scientists are supposed to inquire into what is *true*. Their job is not to worry about religious implications. In so far as the world's religions agree that life's ultimate causes are supernatural, and not purely materialistic, then a similar realization, arrived at by modern, *non-religious* means, confirms their ancient insight, but that's as far as it goes.'

'Modern evidences for intelligence and design in nature can, it is true, confirm *some* of humanity's ancient intuitions about the supernatural but they don't support any particular set or sets of religious claims.

'It's not possible to leap from a reasonable inference of intelligence and design – based on the incredibly clever functionality visible in living things – to endorsing a particular religion, its beliefs about heaven and hell and its requirements for salvation. All the well known modern advocates of ID, like Michael Behe, Stephen Myer and William Dembski, acknowledge this.

'They simply argue that science can now show us the following: (a) that Life is very ancient. (b) Great changes over time *are* recorded in the fossils, but, as we'll see, they do *not* match Darwin's predictions. (c) Life could *neither arise nor evolve* by mere chance, it's not merely highly improbable, it's demonstrably impossible, and, crucially, (d), there is *no* evidence for any such process. Lastly, (e), it is possible to adduce various compelling modern evidences for intelligence and design in living things, more scientifically and more quantitatively than ever before.'

We walked on by the sparkling stream.

---

[34] Lennox, J. 2009. God's Undertaker: *Has Science Buried God?*. Lion; p. 182, emphasis and text in square brackets added.

6 Darwin's Once Fascinating Idea

'So far, Ali, we've asked whether Life could ever *arise* by pure luck. While, for most of history this idea would have been considered unreasonable, ridiculous even, since Darwin it has been mainstream, "Life is just an accident, which evolved by accident," and the classical view became the minority view.

'Yet, as we've seen, various people, even some well known materialists, like Jacques Monod, have shown that this is not merely unlikely, it's demonstrably impossible. But Darwin, not knowing what we know, believed he could find a *wholly materialistic, chance-based* theory of natural history.

'Like others of his day, he had begun to doubt that all Life was instantly created in just six days, a mere few thousand years ago. A view now called young-earth creationism. Modern geological understanding had shown the earth to be far older than the six thousand years, proposed, in the 1600s, by Bishop Ussher, based on his very literal readings of the timelines in the Bible.

'That seems reasonable enough.' Commented Alisha.

'Yes, but Darwin didn't just switch to old-earth creationism, which allows for life to be extremely ancient, but still argues for intelligent causation. He went much further. He wondered whether any kind of intelligent input was relevant? What if there were no supernatural components to reality at all? If so, life must have first arisen and evolved, from there, by pure chance.

'So, he wondered: *Could* a living organism arise by pure chemical luck? Could it vary? Might the variation reproduce itself and, by this process repeating many times, might an entirely new type of creature arise?

'Could, in this way, a fish gradually transform into an amphibian? Could the burly, land living bear, slowly morph into the wonderful whale, with a blowhole above its head, accurate sonar to guide it, collapsible lungs to enable it to dive to great depths and subject to incredible pressures, up to 10,000 feet (1.2 miles!), and the ability to give birth and to nurse its young in the water?

'This imagined process of random variations arising and surviving, he called NATURAL SELECTION. An idea which sounds quite interesting until we realize that all it is telling us is, "That which survives, survives. Why does it survive? Because it's the fittest. How do we know it's the fittest? Because it survives. Why does it arise in the first place? Natural selection cannot tell us."

'Interestingly, despite not having our modern knowledge of the astonishing complexity of living things, there were plenty of scientists even in Darwin's own day who were skeptical of his claims. For example, St George Mivart, who, in 1871, published *On the Genesis of Species,* wherein he doubted that natural selection could *initiate* biological structures. According to S. J. Gould,

'Darwin offered strong, if grudging, praise and took Mivart far more seriously than any other critic... Mivart gathered, and illustrated "with admirable art and force" (Darwin's words), all objections to the theory of natural selection. . . "a formidable array" (Darwin's words again). Yet one particular theme, urged with special attention by Mivart, stood out as the centerpiece of his criticism. ... No other criticism seems so troubling, so obviously and evidently "right" (against a Darwinian claim that seems intuitively paradoxical and improbable).' [Mivart called his criticism] *"The Incompetency of 'Natural Selection' to account for the Incipient Stages of Useful Structures."* [35]

'What did he mean by that?' Wondered Alisha.

'That natural selection cannot initiate living things because, in the aimless nature of materialist thought, there's no reason to select anything at all in the first place, let alone the highly specific combinations of chemicals required.

'Darwin based his concept of macro-evolution on his knowledge of the changes intelligent people, with intelligent aims in mind, could bring to an existing species. Breeders who already had something extremely functional, and extremely clever, like wolves, cows or pigeons to select from. Whereas, Darwinian evolution relies on pure luck for all its alleged creativity.

'Yet, with just a little thought, we can see there is no reason for such an evolution even to *begin* let alone to have any progressive quality. So, if Living things get ever more complex and sophisticated, as Darwin hypothesized, it is just an astonishing fluke. I find it hard to understand how anyone can find this way of thinking intellectually fulfilling. It seems odd to me, Alisha, how so few in our current science culture seem to be interested in more subtle and more interesting possibilities than pure (stupid) chance + mindless matter + mindless natural laws as Life and Mind's supreme creative principles!

'That Life could, it is easy to demonstrate, neither (a) *arise* nor (b) *evolve* by pure luck and that (c) there's *no evidence for any such process,* seems to bother all too few in science today. But, despite this intellectual short-sightedness, what Darwin called natural selection can, at best, only account for the survival and the variation, *within strict genetic limits,* of what *already* exists, it cannot account for its arrival in the first place.

---

[35] Gould, S. J. 1985. "Not Necessarily a Wing" *Natural History,* October, pp. 12, 13. Text in square brackets added.

'In *Darwinism: The Refutation of a Myth,* Soren Lovtrup wrote, 'some critics turned against Darwin's teachings for religious reasons, but they were a minority; most of his opponents ... argued on a completely scientific basis.' He added, prefiguring Michael Behe's theory of irreducible complexity:

'... the reasons for rejecting Darwin's proposal were many, but first of all that many innovations cannot possibly come into existence through the accumulation of many small steps... because incipient and intermediate stages are not advantageous [they provide no selective benefits].'[36]

'Michael Flannery writes that, unlike Darwin's many admirers today,

'By and large, the scientists of his day were *not much impressed* with Darwin's theory. John Herschel called natural selection "the law of higgledy-piggledy," and William Whewell thought the theory consisted of *"speculations" that were "quite unproved by facts,"* so much so that he refused to put the book on the shelves of the Trinity College Library.

Rather, it was the reading elite of London that was captivated with Darwin's theory. "Freethinking" bohemians and assorted society trendsetters grabbed up copies of the *Origin* and later *Descent*. The secular [materialist] creation myth they had long been looking for was finally in hand. ...

In the end, it was Darwin's rhetorical salesmanship that won the day. It's not so much, as some have claimed, that evolutionism was already "in the air" in Victorian England. ... This wasn't achieved by Darwin's prodigious scholarship. It was accomplished by his sheer presentation. It was a pitch easily made because it "sold" a product the intellectual elites had long been waiting for, a theory of life in which God [the super natural, intelligent causation] was superfluous and irrelevant. As everyone in retail knows, you've got to have the right product at the right time. Darwin had both.' [37]

'Remember, so called **natural selection cannot build anything functional from nothing.** Just, when conditions vary, those members of an *existing,* adready-very-cleverly-put-together species born with the mix of genes, correctly copied or slightly mutated, best suited to the prevailing conditions do better than the others. This differential survival of "the fittest," of those best suited to the prevailing conditions, **can only maintain or refine what already exists. It cannot create anything new from scratch**. It has no reason to.

[36] Lovtrup, S. 1987. *Darwinism: The Refutation of a Myth.* Croom Helm Ltd., Beckingham, Kent, p. 275, text in square brackets added.

[37] Michael Flannery. 'Was Darwin a Scholar or a Pitchman?' *Evolution News*, October 20, 2015.

'But, despite the obvious limitations of natural selection, Darwin speculated that a random change in an existing creature *might* arise, it *might* survive and it *might* contribute to an "inconceivably great" (his words) number of transitional versions of the original animal, leading, eventually, to an entirely new creature.

'It seems there was a lot of speculation on Darwin's part?' Mused Alisha.

'Yes, Darwin's theory always consisted in *many* unproven speculations. So, he speculated, might a single-cell microbe, through a vast departure from its original form, slowly morph into a multi-celled worm or might a worm slowly turn into a snail, spider or fish? Or, might a fish gradually morph into a frog: Fish-Frish-Frosh-Frogh-Frog? This idea is known as Darwinian macro-evolution and Darwin hoped the fossils would, eventually, prove him right.

'But, aside from the fossil evidence which, shockingly, has never backed up his ideas, one of the many downsides to his approach was that it had nothing useful to say about the *inner Being, the Mind and the Soul, the Subjectivity and the Sentiency* in living things.

'Rather, his was a theory relating solely to the form sides of life, to MATTER AND MECHANISM only. It was all about the beautiful body of Nature, it had nothing useful to say about her *Subjective or Soul* elements. Nothing about feelings and emotions or meaning and purpose. Odd, because it was an attempt of Darwin's *intelligent mind* and his sensitive *Soul* to make sense of Nature without there needing to be *any intelligence or Soul* in Nature.

'Intriguingly, by contrast with Darwin, the yogi-adepts of India, who are old-earth creationists in their spiritual vision, have always maintained that, over the eons, Life on Earth has waxed and waned across vast life cycles,[38] called Yugas, some of which successive Life Waves may correspond to the mysteriously abrupt emergences of the species seen in the fossil records.

'According to the esoteric perspective of the yogis, the fascinating history of Life on Earth is not primarily an evolution of outer forms but it is a mysterious unfoldment of *sentient interiorities,* of diverse *Soul* states for the many different kinds of beings. For example, the amazing world of the dinosaurs had its own, mysterious *Soul* feel, its own *psychic* signature. It came, it served it's mysterious purpose, and it went.

'Darwin, by contrast, argued for a purely physical unfoldment, in the shape of a tree, with a universal ancestor microbe at its base. A microbe which might vary so much as to give birth to entirely new plants and animals as the first branches of the tree. Might such variants also vary, again and again, as more and more offshoots, until, eventually, at the tips of the branches, would be

---

[38] Davidson, J. 1992. *Natural Creation or Natural Selection? – A Complete New theory of Evolution.* Element

found all today's species? It was an intriguing idea. Only, the fossils have never supported this key idea of one creature gradually changing into another,

"by differences not greater than we see between the natural and domestic varieties of the same species at the present day. ... [and that] **the number of intermediate and transitional links,** *between all living and extinct species,* **must have been inconceivably great.** *But assuredly,* **if this theory be true, such have lived upon the earth.** *"* [39]

'It came as a great surprise to me, Ali, as I began to research more deeply into this controversial subject, to find that, despite all the claims made for Darwin, the fossils disagree with him! Not only is there no "inconceivably great" "number of intermediate and transitional links, between all living and extinct species," his predicted transitional species are never to be found!

'In his recent book, *Darwin's House of Cards,* veteran Darwin observer, Tom Bethell, wrote that he once toured the exhibits at the Natural History Museum in London with its then senior paleontologist, Colin Patterson. Patterson explained to Bethell "that he was looking for cases where the hypothetical common ancestor of two species was identified in the diagrams on display. These would be at the "nodes" [branching points] in the tree of life." But all the hypothetical nodes shown on the diagrams were EMPTY.

'The presumed shared ancestors, of later species which had, according to Darwin, descended and diverged from them, were not depicted because they are **never to be found.** Patterson told Bethel that, "as far as he could see [the] nodes [the places where Darwin's hypothetical common ancestors ought to be] are always empty in the tree of life," and he doubted that they'd ever be filled.

'What many people don't realize, Alisha, is that, despite all the claims, the *fossils refute Darwin.* When new species arose in the deep past, by contrast with what we have so often been told, *they appeared abruptly and fully formed, in a mysterious pattern of big and little bang emergences* with no genuine ancestor species to be found: "in any local area, a species does not arise gradually by the steady transformation of its ancestors; *it appears all at once and fully formed.* " [40] So wrote famous paleontologist, Stephen Jay Gould.

'Much earlier than Gould, another renowned paleontologist, George Gaylord Simpson, stated,

'This is true of all thirty-two orders of mammals . . . The earliest and most primitive known members of every order [when they are first found in

---

[39] C. Darwin, *The Origin Of Species,* Chapter X, "On *the Imperfection of the Geological Record.* Emphasis added.

[40] Gould, S. J. 1980. *The Panda's Thumb.* p. 181-182)

the fossil records] already have the basic ordinal characters, and in no case is an approximately continuous sequence from one order to another known. In most cases *the break is so sharp and the gap so large* that the origin of the order is speculative and much disputed . . .'

'This regular *absence of [Darwin's] transitional [fossil] forms* is not confined to mammals, but is an almost universal phenomenon, as has long been noted by paleontologists. It is true of almost all classes of animals, both vertebrate and invertebrate ... it is true of the classes, and of the major animal phyla, and it is apparently also true of analogous categories of plants.'[41]

'Are you saying that there is no reliable evidence for the very slow, continual evolution which Darwin predicted?' Asked Alisha, looking surprised.

'Yes, I am. The *fossils do not match* Darwin's claims. This is one of the odd things about this debate. While I had been skeptical for some time that any wholly materialistic theory, such as Darwin's, could explain Life, I did not realize, until I started to look a little deeper, just how much the facts of natural history, and especially the fossil records, did not bear out his ideas at all!

'Yet, how many, today, are aware of this, Ali? Very few, I'd say. Why not? Because we're not told. In *Darwin's Enigma,* Luther Sunderland wrote:

'Geology and paleontology held great expectations for Charles Darwin, although in 1859 he admitted that they presented the *strongest single evidence against his theory.* Fossils were a perplexing puzzlement to him because they did not reveal any evidence of a gradual and continuous evolution of life from a common ancestor, proof which he needed to support his theory. Although fossils were an enigma to Darwin, *he ignored the problem* and found comfort in *the faith* that future explorations would reverse the situation and ultimately prove his theory correct. He stated in his book, *The Origin of Species,*

"The geological record is extremely imperfect and this fact will to a large extent explain why we do not find intermediate varieties, connecting together all the extinct and existing forms of life by the finest graduated steps. He who rejects these views on the nature of the geological record, *will rightly reject my whole theory.*"[42]

'Now, after over 120 years of the most extensive and painstaking geological exploration of every continent and ocean bottom, the picture is infinitely more vivid and complete than it was in 1859. Formations have been discovered

[41] Simpson, G.G. 1944. *Tempo and Mode in Evolution.* Columbia University Press; pp. 105, 107. Emphasis and text in square brackets added.

[42] Darwin, C. 1859. *On the Origin of Species by Means of Natural Selection or the Preservation of Favoured Races in the Struggle for Life,* reprint of 6th edition (John Murray, 1902), p. 341–342. Emphasis added.

containing hundreds of billions of fossils and our museums now are filled with over 100 million fossils of 250,000 different species. The availability of this profusion of hard scientific data should permit objective investigators to determine if Darwin was on the right track.

'What is the picture that the fossils have given us? Do they reveal a continuous progression connecting all organisms to a common ancestor? With every geological formation explored and every fossil classified it has become apparent that these, the only *direct* scientific evidences relating to the history of life, still *do not provide any [of the] evidence for which Darwin so fervently longed. The gaps between major groups of organisms have been growing even wider and more undeniable. They can no longer be ignored* or rationalized away with appeals to the imperfection of the fossil record. ...

'By the 1970s prominent scientists in the world's greatest fossil museums were coming to grips with the so-called "gaps" in the fossil record [the absence of the "inconceivably great" number of transitional species Darwin's theory calls for] and were cautiously beginning to present new theories of evolution that might explain the *severe conflicts between neo-Darwinian theory and the hard facts of paleontology.* Back in 1940, Dr. Richard B. Goldschmidt had faced the horns of this dilemma-of-the-gaps with his *hopeful monster theory,* the idea that every once in a while an offspring was produced that was a [gap-jumping] monster grossly different from its parents.[43] Goldschmidt's revolutionary ideas were ridiculed for many years . . . [However, in the 1970s] Niles Eldredge, curator of Invertebrate Paleontology at the American Museum, was collaborating with Dr. Stephen Jay Gould of Harvard, and calling their new theory, aimed at explaining the gaps, "punctuated equilibria."'[44]

'As with Goldschmidt's hopeful monster idea, Eldredge and Gould's notion of punctuated equilibria was an attempt to make sense of the fossils as they actually *are.* They were trying to explain the sudden, *non-evolutionary,* appearances of new species and the complete lack of evidence for the millions of transitional fossils required by Darwin's theory. We'll come back to Gould and Eldredge's notion of punctuated equilibrium in a while.

'In *Natural Creation or Natural Selection?,* biologist John Davidson asks,

'Since such [Darwinian] evolution is still supposed to be in progress, why are there no semi-evolved creatures living now? Why are there NONE FOUND IN THE FOSSIL RECORD? In fact, 300 million-year-old spiders with web-weaving spinnerets have been found in the fossil record, yet

---

[43] Goldschmidt, R.B. 1940. *The Material Basis of Evolution.* Yale University Press; p. 390.

[44] Sunderland, L. 1988. *Darwin's Enigma: Ebbing the Tide of Naturalism.* Emphasis added.

nowhere has any fossilized species been found which is on the way to becoming a spider. It's spiders or something else! All creatures, not just spiders, are well structured wholes.'[45]

'You see, Ali, despite all the claims, it turns out that not only could Life not *arise* by chance, equally devastating for Darwin, there's no fossil evidence that ancient *single-cell* microbes v e r y s l o w l y turned into *multi-celled* worms, trilobites, spiders or fish. There's no evidence that dinosaurs v e r y g r a d u a-l l y turned into birds or mammals. There's no evidence that land mammals, like bears, v e r y s l o w l y turned into deep sea diving, echo-locating whales or that some land based rodent, like a shrew, v e r y g r a d u a l l y turned into fast-flying, upside-down-hanging, sonar guided bats, just the many claims that, "It must be so." Why? "Because Darwin said so." This is not scientific.

'Yet, despite this, in the first edition of the *Origin of Species* Darwin wrote,

'In North America the black bear was seen by Hearne swimming for hours with widely open mouth, thus catching, like a whale, insects in the water. Even in so extreme a case as this, if the supply of insects were constant, and if better adapted competitors did not already exist in the country, I can see no difficulty in a race of bears being rendered, by natural selection, more and more aquatic in their structure and habits, with larger and larger mouths, till a creature was produced as monstrous as a whale.'
Darwin, C. 1859. *On The Origin of Species.* Ch. VI.

'Another of Darwin's fanciful speculations, albeit one which was wisely dropped from later editions. What the *fossils actually show* is that, by contrast with his predictions, new species, when they do arise, appear abruptly and fully formed, in a mysterious, *non-evolutionary,* way, with no discernible ancestors. They stably last, *without changing into anything else,* before, eventually, disappearing into extinction. Then, the next Life Wave of similar or somewhat different species abruptly appears and it goes through the same mysterious process. Remember, 99% of all previous species are now extinct.

'The real-world evidence of natural history has never supported Darwin. Rather, it supports the classical view that more than purely physical processes are going on with living things. As to *how* these sudden materializations of living things took place we are not yet able to say. But, if we allow that *all* of manifested reality, from tiny atom to mighty galaxy, appears from seemingly *no-thing to some-thing, out of the cosmic void – think of the Big Bang –* we

---

[45] Davidson, J. 1992. *Natural Creation or Natural Selection – A Complete New Theory of Evolution.* Element; p. 13–14, emphasis and text in square brackets added.

don't have to go for the kinds of wholly materialistic and wholly mechanical understandings which Darwin and his contemporaries were aiming for.'

'You mean that *all* of nature can be thought of as super natural, because it all emerges in this deeply mysterious way, seemingly out of the cosmic void?'

'Yes. Remember, the most mysterious thing of all is that there is *some-thing* at all rather than *no-thing* at all. How is it possible? Why should there be anything at all, even one atom, rather than nothing? Yet, amazingly, there is.

'What we need to work out is whether it's a *random possible,* as materialists like Darwin and Dawkins argue, or an *intelligent possible,* as philosophical idealists, from Plato to Planck, and all spiritual people, believe.

'As it happens, Darwin was well aware that the fossils did not bear out his theory. He was just reluctant to accept the data. Why? Because he was passionate about his ideas and their potential to give him everlasting scientific glory – if he really had unravelled the mysteries of life. So, he just hoped that the millions of fossilized transitional species required by his theory would, eventually, be found. But, recognizing some of the flaws in his ideas, he wrote,

'LONG before having arrived at this part of my work, a crowd of difficulties will have occurred to the reader. Some of them are so grave that to this day I can never reflect on them without being staggered; but, to the best of my judgment, the greater number are only apparent, and those that are real are not, I think, fatal to my theory.

'These difficulties and objections may be classed under the following heads:- Firstly, **why, if species have descended from other species by insensibly fine gradations, do we not everywhere see innumerable transitional forms?** Why is not all nature in confusion instead of the species being, as we see them, well defined?

'Secondly, is it possible that an animal having, for instance, the structure and habits of a bat, could have been formed by the modification of some animal [like a small rodent] with wholly different habits? Can we believe that natural selection could produce, on the one hand, organs of trifling importance, such as the tail of a giraffe, which serves as a fly-flapper, and, on the other hand, organs of such wonderful structure, as the eye, of which we hardly as yet fully understand the inimitable perfection?

'Thirdly, can instincts be acquired and modified through natural selection? What shall we say to so marvelous an instinct as that which leads the bee to make cells, which have practically anticipated the discoveries of profound mathematicians?' [46]

―――――――――――――

[46] Ibid

'So Darwin had his own doubts?' said Alisha, looking surprised.

'Yes, at times, he had profound doubts. But, as he inclined to the belief that only the physical worlds are real, he had to assume that life arose by mere chance and that there was no purpose or plan or design behind any living thing.

'Now, oddly, Darwin who was obliged by his materialist philosophy to claim that there is *no teleology (purpose) or intelligence* in nature, attributed more or less miraculous powers to natural selection to constantly "improve" (his word) on nature's 'unintelligent' and 'purposeless' designs.

'Isn't that self-contradictory?' Commented Alisha. 'How can something which has no purpose be 'improved'?'

'I agree. Because improvement, by definition, implies purpose and function. But, currently, our science is run by the unreasonable idea that *nothing* in nature has any teleology or purpose. Yet, as soon as something, like a cell, a DNA Code, a heart or a lung *functions* it, inevitably, has purpose.

'Functioning always implies purpose . . . which implies motive . . . which implies desire or love . . . and intelligence. We have to contort ourselves, intuitively, logically and experientially, to claim otherwise. The classical philosophers, like Aristotle, who remarked long ago upon teleology – upon purpose and intelligence in nature – understood this. We seem not to.

'This is why, materialist thinking, its mind so firmly closed to the fascinating intelligence and purposeful functioning visible all around us in nature, has to rely on pure luck to *'unintelligently design'* Life's thousands of correctly sequenced polypeptide chains, the stunning complexity of living cells, the incredible intracellular protein machines, cellular replication, selectively permeable cellular membranes, Life's amazing DNA CODES, the biochemistries of replication, respiration, secretion, excretion, taste, touch, vision and hearing and so many other features of living things.

'This is why the distinguished philosopher, Thomas Nagel, describes the materialist *faith* in mindless chance as a supreme creative force, as *'antecedently unbelievable – a heroic triumph of ideological theory over commonsense...* [which] will come to seem laughable in a generation or two.'[47]

'This unjustified faith that mindless little billiard ball atoms, of carbon, oxygen and hydrogen, 'might,' 'somehow,' randomly bump into each other and 'might,' eventually become *living, breathing, thinking, feeling* Human Beings capable of discovering mathematics and physics, painting the ceiling of the Sistine Chapel, writing Symphonies and creating digital software Codes – all for no reason or point at all. Is this intellectually satisfying?

We walked on.

---

[47] Nagel, T. 2012. *Mind and Cosmos*. Oxford University Press; p.128.

7 Life's Mysterious Explosion

'You know, Alisha, I wonder how many young people, taught to believe in evolution, realize not only that Life on Earth could not *arise* by chance, but that, just as important, it did not unfold at all in the manner predicted by Darwin?'

'Go on...'

'Well, microbial life, it's currently believed, dates back about 3.8 billion years. Yet, during the following 3.2 billion years there is *no evidence* for new creatures slowly emerging out of those ancient microbes, as Darwin theorized.

'Then, around 600–500 million years ago, for reasons no one knows, the mysterious *biological big bang* known as the Cambrian Explosion took place and dozens of new animal forms appeared in a relatively short period, including examples of all the basic anatomical designs known today.

'But, critically for Darwin's ideas, there are *no* plausible Darwinian-style ancestors to be seen in the pre Cambrian strata. The earlier rocks yield *no* evidence that the mysterious creatures of this geological big bang of newly arrived animal life descended from the bacterial (non-nucleated) and, later, eukaryotic (nucleated) cells of the previous 3.2 billion years.'

'So there's nothing *evolutionary* about the Cambrian Explosion?' Said Ali.

'No. The Cambrian reveals the mysteriously sudden appearances of many new Life Forms – *without* plausible ancestor fossils. In *The Naked Emperor: Darwinism exposed,* Dr. Antony Latham discusses how even world famous fossil authority S. J. Gould felt compelled to point out that Darwin's predictions did not match the data. [48] Gould wrote:

> 'Darwin invoked his standard argument to resolve this uncomfortable problem: the fossil record is so imperfect [Darwin alleged] that we do not have the evidence for most of life's history. But even Darwin acknowledged that his favorite ploy was wearing a bit thin in this case. His argument could easily account for a missing stage in a single lineage, but could the agencies of imperfection really obliterate absolutely all evidence for positively every creature during most of life's history? Darwin admitted: 'the case at present must remain inexplicable; and may be truly urged as a valid argument against my views here entertained' *(Origin of Species,* 1859).

---

[48] Latham, A. 2005. *The Naked Emperor: Darwinism Exposed.* London: Janus, p. 39.

'Latham continues, 'As already described, we do know of fossils before the explosion but these *completely overturn* the Darwinian model. For 3.2 billion years we have only microbial life... Then, in a period of about 50 million years: the Ediacaran Fauna which appear unrelated to the Cambrian fauna followed by the host of body plans of the Cambrian explosion.' He quotes again from S. J. Gould's *A Wonderful Life:*

> "Thus, instead of Darwin's gradual rise to mounting complexity, the 100 million years from Ediacara to Burgess may have witnessed 3 radically different faunas – the large pancake-flat soft-bodied Ediacaran creatures, the tiny cups and caps of the Tommotion, and finally the modern fauna, culminating in the maximal anatomical range of the Burgess. Nearly 2.5 billion years of prokaryotic cells and nothing else – two-thirds of life's history in stasis at the lowest level of recorded complexity. Another 700 million years of the larger and much more intricate eukaryotic cells, *but no aggregation to multicellular animal life [as Darwin would predict].* Then in the 100 million year wink of a geological eye, three outstandingly different faunas – from Ediacaran, to Tommotion, to Burgess. Since then, more than 500 million years of wonderful stories, triumphs and tragedies, but [in radical conflict with Darwin] not a single new phylum or basic anatomical design added to the Burgess complement.
>
> Step way back, blur the details, and you may *want* to read this sequence as a tale of predictable progress: prokaryotes [non nucleated bacterial cells] first, then eukaryotes [nucleated single cells], then multicellular life [worms, trilobites, snails, fish etc]. But scrutinize the particulars and *the comfortable [Darwinian] story collapses. Why* did life remain at stage 1 for two thirds of its history if complexity offers such benefits? *Why* did the origin of multicellular life proceed as a short pulse through three radically different faunas, rather than as a slow and continuous rise of complexity [as predicted by Darwin]?"[49]

'So, S. J. Gould, a world famous fossil authority, was pointing out that Darwin's theory is not borne out by the fossil evidence?'

'Correct. Darwin's idea was that, first, a single-cell, common ancestor microbe arose, perhaps by chemical coincidence, as he mentioned in his letter to his friend, J. D. Hooker.[50] Then, he speculated, such a microbe might vary so much, and into so many different forms, as to become, eventually, common

---

[49] Gould, S.J. *Wonderful Life. 1990. The Burgess Shale and the Nature of History,* New York: Vintage, pp. 59–66; emphasis and text in square brackets added.

[50] Darwin to Hooker. 2.2.1871.

ancestor to all today's Life Forms, from Roses to Oak Trees, from Sparrows to Eagles, from Butterflies to Blue Whales, all branching out like a vast tree, with all of today's species at the tips of the branches.

'Except, for several billion years, single microbial cells did *not* give birth to multi-cellular creatures. Then, for reasons no one knows, biology's most famous big bang took place and many – not just one or two – new animal body-plans appeared within a relatively short span of each other. But, there's *no fossil evidence* for any kind of evolutionary arc from complex single-celled microbes to even more complex multi-celled creatures.

'Then, since the Cambrian explosion, in complete contradiction of Darwin's forecasts, not a single new basic anatomical design has arisen. The Cambrian explosion overturns Darwin's acorn-to-oak-tree-of-life idea. Also in conflict with his theory, there are *fewer* basic body plans around today, not more.

'What do you mean by body plan?'

'Body plan means basic anatomical design. There are over thirty. Among the most important, *Cnidaria,* including sea anemones, corals and jellyfish, *Mollusca,* snails, slugs, mussels and octopuses, *Annelida,* marine worms, earth worms and leeches, *Arthropoda,* the biggest group, around 3.7 million species, including the classes *Arachnida,* spiders, mites and scorpions, *Insecta,* 1.5 to 3 million species, including ants, bees, beetles, flies and butterflies, and *Crustacea,* shrimps, lobsters and crabs, *Platyhelminthess,* flatworms, parasitic flukes, tapeworms, *Nematoda,* round worms, over 80,000 species, *Echinodermata,* including starfish, sea urchins, sand dollars and sea cucumbers and *Chordata,* the basic design to which vertebrates belong, about 60,000 species including fish, amphibians, reptiles, birds, mammals and people.

'The mysteriously abrupt – *the very non evolutionary* – arrival of many new animal forms in the Cambrian, which is devastating for Darwin's whole theory, is discussed at scholarly length in Stephen Meyer's, *Darwin's Doubt: The Explosive Origin of Animal Life and the Case for Intelligent Design.*[51]

'The 'doubt' in Meyer's title refers to Darwin's own recognition that the animals of the Cambrian explosion had no discernible ancestors and that this could refute his entire theory. Darwin wrote:

'*The difficulty of understanding the absence of vast piles of fossiliferous strata,* which on my theory were no doubt somewhere accumulated before the Silurian (Cambrian) epoch is very great. I allude to the manner in which

---

[51] Meyer, S. C. 2013. *Darwin's Doubt: The Explosive Origin of Animal Life and the Case for Intelligent Design.* Harper One. Book description.

numbers of species of the same group ***suddenly appear*** in the lowest known fossiliferous rocks.'[52]

'Meyer explains how the sudden appearance of the major animal groups in the Cambrian is as discontinuous as the origin of Life itself. Some have argued that the emergence of new creatures at that time only *seems* discontinuous because any precursors to the Cambrian creatures must have been soft-bodied and lacking in bones or shells which could be preserved as fossils.'

'You mean there were pre-Cambrian ancestors but they were not preserved?'

'Yes, but, as Meyer points out, plenty of soft-bodied organisms are well preserved in the rocks from both before, during and after the Cambrian period, so the new animals, appearing so abruptly at that time, are not the descendants of earlier creatures.

'In *Darwin's Doubt* Meyer discusses the mysterious arrival of the many new animal forms in the Cambrian, one of several biological big bangs which tend to refute Darwin's entire theory. Further, he explains, random genetic mutations would not have the creative power to produce these new forms.

'He also discusses the *key* mystery of the origins of the intriguing, DIGITALLY CODED instructions essential to the construction of these new animal forms. The highly complex instructions of which Darwin and his steam era contemporaries knew nothing. Casey Luskin writes,

> 'Meyer, in particular, argues that the [random] mutation and natural selection mechanism lacks the creative power to produce both the genetic and epigenetic information necessary to build the animals that arise in the Cambrian. Meyer offers *five separate lines of evidence* and arguments to support this latter claim. He also later describes and critiques six post neo-Darwinian evolutionary theories and makes a positive argument for intelligent design. . . .
>
> First, Meyer argues that the neo-Darwinian mechanism cannot efficiently search combinatorial sequence space to find the exceedingly rare DNA sequences that yield functional genes and proteins.
>
> Then he cites multiple peer-reviewed studies showing that *multiple coordinated mutations* would be necessary to produce functional proteins, but these could *not* arise within realistic waiting times allowed by the fossil record. ... [This is] *devastating* to Darwinian explanations of the origin of complex features.

---

[52] Darwin, C. 1996. *On the Origin of Species.* Oxford University Press, p. 249.

... the neo-Darwinian mechanism could *never* produce new body plans given that mutagenesis experiments show how early acting body plan mutations – the very mutations that would be necessary to produce whole new animals from a pre-existing animal body plan – ***inevitably produce embryonic lethals*** [they are FATAL to the organism]. . . .

Finally, Meyer raises the problem of the origin of epigenetic (i.e., "beyond the gene") information necessary to build new animal body plans, a problem that has led many evolutionary biologists to seek a new theory of and mechanism for major evolutionary innovation.

Meyer also looks at various "post-Darwinian" models (e.g., evo devo, punc eq, [punctuated equilibria], self-organization, neutral evolution, etc.) and shows why they too [all] fail to explain the origin of [the] information necessary to generate new body plans.'

'Dr Meyer, Luskin concludes, successfully establishes *intelligence,* not chance, 'as the only known cause capable of generating the INFORMATION and top-down design that are required to build the animal body plans which appear explosively in the Cambrian period.'[53]

'Satisfied by Meyer's arguments that the Cambrian Big Bang is real, the animals of that time really did appear out of the blue, this reviewer writes:

'Essentially, getting "order" from natural self-organizing process[es] and getting "information" are two totally different things. "Order" is easy . . . [e.g. crystals]. But the very nature of information, whether in DNA or human writing, precludes natural [purely material] forces from generating it. DNA can hold information precisely because [as with our own languages] there is no *natural* force demanding the nucleic acids [the DNA alphabet letters, ACT&G] be in one location or another. All information requires this type of "contingency" [physical indeterminacy], that is, openness to many possible choices; a system which is driven to one required state [by the regular ordering of regular physical laws] holds no [instructional] information.'[54]

'So Snoke's point is that while languages and software can be written in many different kinds of ink, including deoxyribonucleic acid (DNA) ink, and stored in many different kinds of media – bio-chemical, magnetic, optical – they cannot be created by them?' Wondered Alisha.

'Correct. They can only come from intelligent sources. Randomly tumbling digital coding symbols around will never generate anything meaningful.

---

[53] *Darwin's Doubt,* amazon review, C. Luskin. Emphasis added.

[54] *Darwin's Doubt,* amazon review, D. Snoke.

'You know, Ali, after I began to have my doubts about Darwin's ideas, and to do more research on them, I was shocked to discover just how much his ideas conflicted with some of the most basic facts of natural history.

'For example, I was surprised to find out that, by contrast with what we have been led to believe, the species arose suddenly and fully formed, most dramatically in the Cambrian, *not* gradually, as Darwin predicted. He imagined that earlier, minor variations in Life's designs would precede later, greater variations. The fossils reveal the opposite. Big differences in the basic animal body plans arrived all together in the Cambrian.

'I was also struck by Art Battson's observation that the conspicuous *stability* of the species over time 'suggests the existence of natural processes which [actively] *prevent* major change.'[55] As Battson says, had Darwin not been so attached to his theory, he might have tried to work out, firstly, why the species appear in such an *abrupt* and non-evolutionary way and, secondly, why they do *not* gradually morph into new animal shapes, as he predicted, but remain extremely stable.

'Unlike Darwin, we also now know that cells deploy extremely sophisticated techniques to ensure that they precisely do not randomly mutate in the way his followers maintain. They work, very actively, to *prevent* random genetic copying errors (mutations).

'For 150 years it's been claimed that Darwin successfully explained the origins of the species but that, unfortunately, there's just not enough fossil evidence to confirm his theory. Battson reflects that this "puts Darwin's theory in the curious position of explaining the data we *don't* have while ignoring the data we *do*," the stability of the basic body plans and the stark fact that "NO TRANSITIONAL SPECIES [dead or living] showing evolution in progress between two [macro-evolutionary] stages HAS EVER BEEN FOUND."[56]

'So, given that the Cambrian data contradicts Darwin and the species do *not* morph into new forms over time,' began Alisha, 'why don't scientists develop theories to explain the long-term *stability* and the natural *limits* to biological change,[57] rather than trying to force the evidence to fit Darwin's ideas?

'Because, if science now accepts that most of Life's basic designs arrived in *one go,* without ancestors, and that the subsequent coming and going of the species is similar to the way our own designs appear – each new or modified

---

[55] Battson, A. *On the Origin of Stasis by Means of Natural Processes.* Battson's website. Emphasis and text in square brackets added.

[56] Milton, R. 1992. *Shattering the Myths of Darwinism,* Inner Traditions, back cover, emphasis and text in square brackets added.

[57] Battson's interesting suggestion regarding what we should be looking at in natural history.

model abruptly appearing, stably lasting, without morphing into anything else, its view of natural history as a wholly unintelligent process is refuted.

'But isn't it odd that the very fossils, which are supposed to be so central to Darwin's ideas, have never supported him?' Mused Alisha.

'It is odd. But, once many of us accept a theory, especially one leading to a massive change in worldview, it can be very hard to persuade ourselves that we may have been mistaken. It's like trying to turn a big ship around. Darwin's theory predicts slow, non-jumping change, but, as Dr. Antony Latham says:

> 'Check any textbook on the subject and you will see that *sudden appearance and discontinuity* is the norm. The problem that Darwin so clearly saw has not gone away with time and the fossil record seems now to REFUTE the entire basis of his theory of gradual changes.'[58]

'As Battson reflects,

> 'Darwin's general theory of evolution may, in the final analysis, be little more than an *unwarranted extrapolation* from microevolution [finch beaks, peppered moths] based more upon [materialist] philosophy than fact. The problem is that *Darwinism continues to distort natural science.*' [59]

'The trouble is, while Darwin correctly observed that species can vary to a *limited* extent, as can be seen with finch beaks and peppered moths, his core belief that an existing animal could change into an entirely new type due to essentially random processes is completely unsupported: fish to amphibian, dinosaur to bird, bear to whale, rodent to ape, ape to human.

'At one time, his ideas might have sounded more scientific than what had gone before, but, despite all the claims, the data has never supported them. In the *Origin of Species* Darwin initially wrote,

> 'In North America the black bear was seen by Hearne swimming for hours with widely open mouth, thus catching, like a whale, insects in the water. Even in so extreme a case as this, ... I can see no difficulty in a race of bears being rendered, by natural selection, more and more aquatic in their structure and habits, with larger and larger mouths, till a creature was produced as monstrous as a whale.'[60]

---

[58] Latham, A. *The Naked Emperor: Darwinism Exposed.* London: Janus, p. 61 emphasis and words in square brackets added.

[59] Battson, A. 1997. *Facts, Fossils, and Philosophy.* Author's website. Emphasis and text in square brackets added.

[60] Darwin, C. 1859. *On The Origin of Species.* Ch.VI.

'But the fossils record no such fantastical transformations. No, new species appear, abruptly, with no obvious predecessors. They stably endure, without changing into anything else, and, eventually, they disappear into extinction.

'Then, the next mysterious Life Wave of similar or somewhat different species intriguingly appears. It is stable, it doesn't give birth to major new forms, as Darwin predicted, if his theory be true, and, eventually, it, too, ends. Remember, 99% of previous species are now extinct.

'Darwin was aware that the fossils did not really support his ideas but he was reluctant to accept this. He just hoped that the millions of fossilized, transitional creatures gradually turning from one kind of animal into a different kind of animal would, eventually, be found. But, he conceded,

> 'Firstly, why, if species have descended from other species by insensibly fine gradations, do we not everywhere see *innumerable transitional forms?* Why is not all nature in confusion instead of the species being, as we see them, well defined? [Separate and in discrete kinds.] Secondly, is it possible that an animal having, for instance, the structure and habits of a bat, could have been formed by the modification of some animal [like a shrew] with wholly different habits?'

'Darwin was right to have his doubts because, it's now clear, his key idea that one creature might gradually morph into another was mistaken. *There is no evidence for such transformations, let a lone a vast number of them.* The species appear abruptly, they are stable, then, mostly, they die out, to be replaced, in due course, by fresh Life Waves of similar or different species.

'Like many of the things we make, they are connected by design. For example, the shared skeletal patterns found in many vertebrates, and the many shared instincts and soul qualities.

'It is easy, for example, to see that our Cars are *conceptually* connected as we work through the layers of 'fossil vehicles' in a junk yard, modern Fords all the way back to Model T. They share many design similarities and there is a *succession,* but it is one driven by intelligence, not by blind chance.

'But don't the fossils show some continuities of form, for example between fish and land creatures?' Asked Ali.

'Yes, but continuities of form are not evidence for 'random' or 'unintelligent' design. Many of our designs show continuities of form. In *The Naked Emperor: Darwinism Exposed,* Dr. Antony Latham writes:

> 'By looking in detail at the known fossils of the earliest tetrapods we can definitely see continuity between the lobe-finned precursors and those first land walkers. Any creationist must take this into account. When we look at

either Acanthostega or Ichthyostega we see the following features which definitely show that they are based on a fish anatomy: they retain a sort of fish shape, they apparently retained internal gills, they retained a lateral sensory line on the body (as in fish) and they have a sort of tail fin – albeit of a different shape. Here is continuity amidst the discontinuity of the other new features such as the limbs. We see here that, even though there is *an unexplained leap* to terrestrial locomotion, there is linkage with fish . . . such continuity is common in different groups of animals. This does *not* validate Darwinism, however. These are *not* " transitional" in the sense that Darwin meant and longed for. We see too much that is *suddenly new* to call the first tetrapods [four legs] transitional. There is ***no gradual evolution here*** but we do see earlier forms being a sort of template for the creation of later forms. The appearance of tetrapods is *just one* of the saltations (leaps of form) that characterize the fossil record.'[61]

'Latham expresses the wish that,

'I hope that I will have demonstrated that there are indeed *totally unexplained yawning gaps* in the fossil record but that there is also continuity between succeeding forms hence the fishlike characteristics of the first tetrapods. They appear with all the attributes of land walkers, *suddenly* – but retain signs of their lineage. Darwinism requires smooth continuity always. We do not see this. There have been enough fossil beds examined (particularly in Greenland) for a clear picture of the fish to tetrapod [four legged] evidence. The transition to tetrapod is ***sudden***. Darwinists will put this down to inadequate fossil preservation but this argument is now wearing thin. The challenge to the Darwinist is to *show us even one case in the history of life where a macro change has occurred smoothly with well-defined, gradually transitional intermediate forms*. Darwin waited [for his predictions to be proven true] and we still wait.'[62]

'As students we were taught that the fossils provide all the evidence we need for Darwin's theory yet, as Geoffrey Simmons points out,[63] nowhere are there any half-fish-to-half-salamanders or part-ape-to-part-human transitions to

---

[61] Latham, A. *The Naked Emperor: Darwinism Exposed.* London: Janus, p. 76, emphasis and words in square brackets added.

[62] Ibid, p. 77 emphasis and words in square brackets added.

[63] https://evolutionnews.org/2018/08/inexplicable-species-and-the-theory-of-evolution/

be found. Darwin said there should be an "inconceivably great" number of transitional species on the way to becoming something else, if his theory be true, but the *numberless transitional species he called for are never to be found.*

Not only this, but researchers now realize it would take millions of biochemical changes "for dinosaurs to evolve into birds, flat plants into trees, fish into amphibians," [64] not just a few random genetic mutations. Simmons writes,

> 'Darwin wrote that whales came about as a result of bears going to sea. ... [Yet] No one knows how blow holes came about . . . or how the internal lungs became connected up to these holes in a way that prevents drowning. Or, how a massive communication center, found in their heads, came about. Or, how the ability to depressurize body segments during deep dives evolved. ...
>
> Whales are not the only misfit to smooth transitions, just the largest. ... [Other] Standouts are kangaroos, woodpeckers, platypuses, giraffes, butterflies, octopuses, skunks, bombardier beetles, the red tide, dolphins, fireflies, tardigrades, sloths, and all micro-organisms. Maybe viruses, too.
>
> ... Just breeding a horse into a faster horse doesn't eventually change it into something fast like a cheetah. It's simply a faster horse. ... It's true, natural selection does happen in a variety of situations, but *it doesn't change one species into another.*'[65]

'Art Battson notes that Michael Ruse, a well known philosopher of science, once said that modern evolutionary theory is like a 'secular religion,' where,

> 'evolution as a scientific theory makes a commitment to a kind of naturalism [materialism], namely, that at some level one is going to exclude miracles [supernaturalism] and these sorts of things *come what may* [i.e. regardless of evidence or alternative hypotheses like intelligent design].' Ruse,1993.[66]

'The problem with this dogmatic anti-supernaturalist attitude in science, Ali, which has now become *its own kind of faith,* is that it excludes, from the first, all theories of intelligent, non-physical causation. Which, if you think about it, makes us, and everything we see around us, meaningless and accidental. Materialism drains the world of all intelligence and all meaning, then, peculiarly, it asks us to be intellectually satisfied with this illogical fare.'

---

[64] Ibid

[65] Ibid

[66] Battson, A. 1997. *Facts, Fossils, and Philosophy.* Author's website. Emphasis and text in square brackets added.

8 Pre-Life Soup: Myth and Reality

'Here's another myth, relating, again, to the very beginnings of Life. Now, for the materialist belief system to be true, we *have* to accept the idea that Life arose by pure chance. This is to put it simply, but, at root, it *is* the claim. There is no avoiding the fact that materialist thinkers like Darwin are claiming that living things are, at base, entirely 'unintelligently designed' and that it's all just a cosmic fluke. No intelligence was or is ever involved.

'Now I wonder, Ali, if you remember the images of heaving, pre-life chemical-soup seas and the thundering skies depicted in some children's science books and older TV programs? Followed by the claim, 'Hey presto, Life!'? Well, it may surprise you to know, there is *no evidence at all* for any kind of pre-life chemical soup. Molecular biologist, Michael Denton, writes,

> 'Rocks of great antiquity have been examined over the past two decades and in none of them has any trace of abiotically produced organic compounds been found . . . Considering the way the *pre-biotic [pre-life] soup* is referred to in so many discussions of the origin of life as an already established reality, it comes as something of a shock to realize that there is absolutely *no* positive evidence for its existence.'[67]

'Secondly, even if there *was* a pre-life soup, the kinds of chemical reactions which produced a few basic amino acids in the Miller–Urey experiments are reversible. This means that the forces giving rise to such basic building blocks will destroy them as quickly as they are formed. Miller and Urey solved this problem by *intelligently designing* a chemical trap to save their blocks. Such a trap, of course, would not exist in the mindless nature of materialist thinking.

'Thirdly, Eastman and Missler write, 'The destructive effect of oxygen, ultraviolet radiation from the sun ... makes it unlikely that significant quantities of viable nucleotides and amino acids could ever accumulate in the primitive ocean. However,' they explain, 'even if they did accumulate in sufficient quantities, the next step is to explain how they combined to form the self-duplicating DNA molecule and the thousands of proteins found in the simplest living cells. For the materialistic scenario to be taken seriously, it must provide

---

[67] Denton, M. 1985. *Evolution: A Theory in Crisis.* Burnett Books; p. 261, emphasis added.

a plausible explanation for the origin of these enormous molecules without the introduction of biochemical knowhow...' They continue,

'One of the most difficult problems for the materialistic scenario on the origin of life is something called *molecular chirality*. The building blocks of DNA and proteins are molecules which can exist in both right and left-handed mirror-image forms. This [left and right] "handedness" is called "chirality." ...

'In all living systems the building blocks of the DNA and RNA exist exclusively in the *right-handed* form, while the amino acids in virtually all proteins in living systems, with very rare exceptions, occur only in the *left-handed* form.

'The dilemma for materialists is that *all* "spark and soup-like" experiments produce a mixture of 50% left (levo) and 50% right- handed (dextro) products. ... Unfortunately, such mixtures are *completely useless* for the spontaneous generation of life.

'For 80 years chemists have been trying to synthesize optically pure mixtures of amino acids in the lab using stochastic [random] chemical processes. However, this has never been accomplished. According to physical chemists, IT IS IMPOSSIBLE . . . Miller and Urey acknowledged that the chemical makeup of their experiment consisted of equal portions of left-handed and right-handed amino acids.'

'And,

'Complex molecules such as DNA and proteins are built by adding one building block at a time onto an ever-growing chain. In a "primordial soup" made up of equal proportions of right and left-handed building blocks, there is an equal probability at each step of adding either a right or left-handed building block.[68]

'Consequently, it is a MATHEMATICAL ABSURDITY to propose that *only* right-handed nucleotides would be added time after time without a single left-handed one being added to a growing DNA molecule. Sooner or later an incorrect, left-handed nucleotide will be added. The same goes for proteins. . . .

---

[68] 'In practice, laboratory experiments have shown that right-handed building blocks have a slightly greater affinity, or attraction, for other right-handed building blocks. Therefore, at each step in the addition of another building block, there is a 3/7 chance that the next one added will be the same optical isomer as the one previously added. . . . The smallest known free living life forms, bacteria, have about 12,000,000 nucleotides in their DNA. If we were to calculate the odds of adding twelve million successive right-handed nucleotides to the growing chain, without a single left- handed one being added, it would be 5 raised to the 12 millionth power, (5/12,000,000)!' Eastman & Missler

'Consequently, if even one nucleotide or amino acid with the incorrect "handedness" is inserted into a DNA or protein molecule, the three-dimensional structure will be annihilated and it will cease to function normally.'[69]

Alisha looked puzzled. She said, 'Ok, if people are showing, once again, that 'life-by-accident' is impossible, it is literally a *mathematical absurdity,* how come anyone still believes it?'

'For philosophical and emotional and political reasons.'

'But it doesn't work. . .' Alisha said, looking baffled.

'Yes, but, as Frederick Hoyle pointed out, the debates about Life's origins are driven as much by emotion as by reason. Nature's deeper causes cannot be physically seen. They can only be deduced using logic, sensed intuitively or experienced by extrasensory means.[70] But our current science accepts none of these in relation to this issue. As to extra sensory perception, all psychic and spiritual experiences and perceptions, even if they are well corroborated, tend to be regarded as hallucinations, hoaxes or delusions. Way back in 1954 George Wald, a Harvard biochemist and Nobel laureate, wrote:

'One has to only contemplate the magnitude of this task to concede that the SPONTANEOUS GENERATION of a living organism is IMPOSSIBLE. Yet we are here-as a result, I believe, of spontaneous generation.'[71]

'That's odd,' said Alisha, 'it was clear that life could not possibly arise by chance but it 'must have.' That's not very scientific?'

'No, but as Sir Frederick Hoyle noted, it's a psychological stance flowing from a dislike of the philosophical alternatives to materialism.'

'Ok, Ollie, the idea that Life could just be chemical coincidence clearly has *no scientifically justifiable basis.* Yet so many people today do believe that 'science has shown' Life to be just a lucky accident which evolved by accident. Which, if true, means that *this* life is all there is, the world's spiritual traditions are all nonsense, and there's no need to appeal to anything beyond the known laws of chemistry + pure luck to explain anything. How do you answer them?'

'That's a great question, but it's not an easy one to answer, because the debates around these issues can be so contentious. For example, from the

---

[69] Eastman, M & C. Missler. 1995. *The Creator Beyond Time and Space.* Emphasis added.

[70] Steiner, R. 2008 (written 1904–5). *Knowledge of the Higher Worlds and its Attainment.* Tompkins, P. 1997. *The Secret Life of Nature,* Thorsons

[71] Quoted from George Wald, 'The Origin of Life', *Scientific American* 191:48 (May 1954). Emphasis and text in square brackets added.

materialist point of view, it's not just a matter of conceding that the *super natural* and the spiritual dimensions of reality are real and we can all move on.

'No, not only is there no belief in the spiritual, there is often a deep anger towards Life, because of its many challenges and difficulties, and there can be considerable hostility towards the world's religions as well.'

'Why towards the world's religions?'

'Because, regardless of what their founders may have originally intended, they have often been responsible for so much conflict and suffering. Horrible religious wars have been fought, and, for centuries, priests and theologians used their authority to oppress people and to prevent the spread of better ideas.

'If you dared to question them, or to say anything they objected to, you could be accused of heresy or blasphemy and killed. In some places this still happens. This makes it difficult for some to consider any evidence which shows spiritualism to be *true* and materialism to be *false*.'

'Others, like Vicki, an old friend of mine, think that a scientific explanation *has* to preclude a supernaturalist explanation. She seems to find it hard to understand that it's the job of science to discover out what is *true,* not to cling to a particular ontology like materialism. But, like many today, she confuses science with a materialist ontology and she assumes materialism to be true.

'This is why she cannot get her head around the modern arguments in favor of intelligent causation, or what some call intelligent design. She believes, *"Science has proved that Life started by pure chance and the world's spiritual traditions are all make believe, even if they still have some cultural merits."*'

'Hmm...' said Alisha, thoughtfully. 'It's quite an intellectual impasse then.'

'Yes. Right now, there is an intellectual battle between those, like Vicki, who believe that all talk of the supernatural and the spiritual is just wishful thinking, – that Life is just an amazing cosmic accident which evolved by accident, that science and the materialist belief system are one and the same and only materialistic explanations of Life can be taken seriously, – and those, currently a small, if gradually growing, minority in science, who argue that Darwin's materialist paradigm does not work and it needs to be questioned.

'For example, modern intelligent design advocate, V. J. Torley, describes how, Professor James M. Tour, *"one of the ten most cited chemists in the world...* [who] has authored or co-authored 489 scientific publications and his name is on 36 patents,"* took the courageous step, along with over 700 other scientists, "back in 2001, of signing the Discovery Institute's *'A Scientific Dissent from Darwinism, "* which reads: "We are skeptical of claims for the ability of random mutation and natural selection to account for the complexity of life. Careful examination of the evidence for Darwinian theory should be encouraged."

'At his website Professor Tour writes:

'Although most scientists leave few stones unturned in their quest to discern mechanisms before wholeheartedly accepting them, when it comes to the often gross extrapolations between observations and conclusions on macroevolution, scientists, it seems to me, permit unhealthy leeway. When hearing such extrapolations in the academy, when will we cry out, *"The emperor has no clothes!"?*

'. . . I simply **do not understand, chemically,** how macroevolution could have happened. Hence, am I not free to join the ranks of the skeptical and to sign such a statement without reprisals from those that disagree with me? Furthermore, when I, a non-conformist, ask proponents for clarification, they get flustered in public and confessional in private wherein they sheepishly confess that **they really don't understand either.** Well, that is all I am saying: I do not understand. But I am saying it publicly as opposed to privately.

'**Does anyone understand** the chemical details behind macroevolution? If so, I would like to sit with that person and be taught, so I invite them to meet with me. Lunch will be my treat. Until then, I will maintain that no chemist understands, hence we are collectively bewildered. And I have not even addressed **origin of first life issues**. For me, that is even more scientifically mysterious than evolution. **Darwin never [formally] addressed origin of life,** and I can see why he did not; he was far too smart for that.

'Present day scientists that expose their thoughts on this become ever so timid when they talk with me privately. **I simply can not understand the source of their confidence when addressing their positions publicly**.'

'Torley describes how, in a talk in 2012, Professor Tour went further and declared that **no scientist that he has spoken to understands evolution** – and that includes Nobel Prize winners. Here's what he said when a student in the audience asked him about evolution:

'. . . I don't understand evolution, and I will confess that to you. Is that OK, for me to say, "I don't understand this"? Is that all right? . . .

'Let me tell you what goes on in the back rooms of science – with National Academy members, with Nobel Prize winners. I have sat with them, and when I get them alone, not in public – because it's a scary thing, if you say what I just said – I say, "Do you understand all of this, where all of this came from, and how this happens?" **Every time that I have sat with people who are synthetic chemists, who understand this, they go "Uh-uh. Nope."** . . .

'I was once brought in by the Dean of the Department, many years ago, and he was a chemist. He was kind of concerned about some things. I said, "Let me ask you something. You're a chemist. Do you understand this? *How do you get DNA without a cell membrane?* [Which is needed to protect the DNA inside the membrane.] *And how do you get a cell membrane without a DNA?* [Which has to CODE for the cell membrane needed to protect it.]

'And how does all this come together from this piece of jelly?" We have no idea, we have no idea. I said, *"Isn't it interesting that you, the Dean of science, and I, the chemistry professor, can talk about this quietly in your office, but we can't go out there and talk about this?"*

'You see, Ali, there is a lot more going on here than pure science. People's beliefs about reality are at stake. Consensus reality is at stake. Professional reputations are at stake. It takes a brave soul, therefore, like a Behe, Wells, Meyer or Tour to put their head above the parapet and risk the ridicule of those whose minds are closed or unable or unwilling to think for themselves.

'Darwin, of course, was not in this category. Not at all. He was going against the prevailing attitudes on origins, and his own dear wife's cherished religious beliefs, which took a lot of courage on his part. But this, by itself, did not make his views true, intriguing as they might once have seemed.

'However, since then, his views, and his materialist philosophy, have become the new mainstream and it now takes those of equal courage, like Professor Tour, to point out the shortcomings of Darwin's materialist beliefs in general and of his theory of macro-evolution in particular.

'Regrettably, currently, those like Tour who point out the flaws in Darwin's ideas are generally not met with an open minded curiosity but with a marked unwillingness even to consider that Darwin may have been mistaken and that the entire materialist paradigm needs to be questioned.

'Unfortunately, many people, like my old friend Vicki, will only consider looking properly at the *actual* evidence – the biochemical, fossil, genetic, embryological, mathematical, probabilistic and informational evidence – in a way which allows for the classical possibility of intelligent or supernatural causation when the winds of fashion and group-think begin to change.

'Then, *when the tipping point comes,* for fear of being thought stupid or of being left behind, they'll quietly shift their positions, denying that they ever advocated anything as ridiculous as the – presently, oh so very fashionable – idea that stunningly complex Life or intelligent Mind could arise by pure stupid chance from mindless, dumb matter and ignore you if you attempt to point out that this was not their position at all, not so very long ago!'

We continued on our way by the gently flowing river.

9 Science and Materialism

After a few moments, Alisha said, 'One thing I don't understand, Oliver, is why is science today so committed to *materialism* and so opposed to all ideas of the non-physical and the spiritual? Isn't it unreasonable to try to explain everything, even our very minds, in terms of pure, dumb, luck? What if materialism isn't true? Shouldn't scientists be interested in this key question?

'Those are all very valid questions, Ali. But the *subtle* causes of things, if real, cannot be seen physically. They can only be intuited or inferred. Also, in the light of modern science, which doesn't readily mesh with the old creation stories, many 19th century scientists and intellectuals, including Darwin, began to lose faith in the traditional religious accounts of things.

'Thirdly, it would be a big shift for our sciences to admit, once more, to the real existence of the spiritual and the supernatural. A shift which many think would be damaging to the practice of science as presently understood.'

'Why?'

'Because, currently, to accept any ideas of intelligent causation into our understandings of nature conflicts with two key science-method conventions, known as methodological materialism and methodological naturalism or MM/MN. According to these conventions, modern scientists, unlike their medieval predecessors, are trained to assume, they are *obliged* to assume, that all things in nature can be explained in purely physical terms.

'For all scientific purposes, they must presume that nature is material alone, that there is no teleology or design at any stage and that random chemical coincidences create all – genes, DNA, hearts, lungs, livers, brains, everything.

'Ok. But what if some things in nature do have intelligent causes? Wouldn't a wholly materialistic methodology make it impossible to explain them?'

'Yes, that's true. Even so, the original idea of disregarding two of Aristotle's famous four causes, those of shaping ideas and purposes and of assuming, for practical, scientific purposes, that only material and mechanical causes are operative, did have some benefits. It did help to steer scientists away from the quaint medieval tendency to resort to supernaturalist explanations whenever it seemed difficult to explain something in physical terms alone.

'This new approach, adopted at the dawn of the modern age by thinkers like Sir Francis Bacon, would, it was thought, reduce the temptation to indulge in what is now called 'god or intelligence of the gaps' thinking, where, if scientists don't understand something, they might be tempted to fill a knowledge gap by saying, "This living thing may be supernatural in origin so we'll never be able to understand it. We must just accept this and move on."'

Alisha looked puzzled. She said, 'That doesn't make sense. Just because something has an *intelligent* cause, it doesn't mean we'll never be able to understand it, do good science on it or harness it for practical purposes.'

'Ok...'

'Well, can't methodological materialism or naturalism lead just as easily to 'unintelligent design' or 'lucky-chemical-flukes' of the gaps thinking?'

'Go on...'

'Where every time scientists don't know how something arises, by extremely involved processes, they attempt to fill their knowledge gaps by saying, "This amazing Living thing *must have* arisen by pure chance. It might *look* as though it is cleverly put together but it's an illusion. It's just another random phenomenon like this randomly evolving river valley here."'

'Ok, Ali, it's true that whether we put forward intelligent or mindless causes to explain something we are attempting to fill a knowledge gap. We just have to decide which type of explanation better fills the gap. A landscape, like this river valley, can, it is true, be explained in terms of unintelligent design. But Cell Phones, or Living Cells, far more complex than anything we make, only make sense in terms of intelligent design or ID. However, despite this, Francis Bacon's MM/MN approach did, in its day, make some good points.'

'I'm not convinced,' said Alisha, looking skeptical.

'Well, take thunder and lightning. For a long time they were thought to be supernatural phenomena. But, eventually, scientists found they could be described, at least partly, in purely *physical* terms. Similarly for magnetism and electricity. If scientists had given up and said, "They're supernatural, they'll always be beyond our understanding," we wouldn't have the many benefits which our modern scientific understanding of them has made possible.

'Hold on, Oliver, just because we better understand electricity, DNA or genetics today, and can use them for practical purposes, it doesn't prove that their ultimate origins lie in something unintelligent or purely chance-based.'

'Ok, it's true that the distinctions we draw between the 'everyday natural' and the 'beyond-the-usual' natural are of our own making. What exists is what exists, be it physical or spiritual, be it obvious or subtle. And, whether we call it natural or *super* natural, we can attempt to investigate *all* of it, using rigorous methodologies – that is scientifically.

'Clearly, it *is* easier to investigate some things than others. But even if we are unable to find out everything about something, it doesn't mean we can discover nothing at all. For example, we can work out, as Darwin failed to do, that neither Life nor Mind could arise by chance alone and that intelligence, whether we call it natural or *super* natural, must, somehow, be involved.

'That we may not yet be able to work out exactly how subtle, intelligent forces give rise to Living, Feeling, Thinking Beings like us does not mean: "They cannot exist then." Just because some things are easier to investigate than others, it doesn't mean we should give up and conclude that, "We cannot explore them at all or that they cannot exist!" That would be lazy and defeatist.

'The issue here is that until not so long ago, if scientists didn't understand something, like electricity or magnetism, they called it *super* natural...'

'Because, they believed it to be *beyond* everyday nature in some way?'

'Yes. But later, when they were able to make more sense of it, they decided it was just 'ordinary' or 'everyday' natural. It wasn't anything special after all.

'Yet the phenomenon hadn't changed. Before, they saw electricity as magical and mysterious. Then, when, to some extent, they had understood it, and harnessed it for practical purposes, instead of continuing to view it as another miracle of a *basic* reality which is, in its very *essence,* miraculous, it was demoted it from the 'super' to the 'everyday,' forgetting that the fact that anything at all exists is the *most extraordinary or most miraculous thing of all!*

'Remember, Ali, nature, all of it, physical and spiritual, must come from something, which *'just is.'* No one can say *why* it 'just is,' or why it gives rise to reality. But we can, surely, try to work out as much as we can about its nature. Is it mindless or intelligent? Is it just physical? Or does it consist also in *Soul and Spirit* elements? Is it mono-dimensional or multi-dimensional?

'The world's prophets and psychics have always said it is *multi-dimensional,* that it consists, also, in subjective elements and that its more subtle dimensions can be known, to some extent, by *inner* means, by shifts in our own consciousnesses as we travel on the various spiritual paths. Apart from this, we can use philosophy and logic to inquire into its likely attributes.

'Having used logic and other elements of modern science to rule out pure chance as the source of Living things, we can, reasonably, conclude that there is more going on with them than mere luck. But, currently, our science doesn't see it this way. As each mystery of physical nature has been solved, even if only partially, it has gone from being thought of as *super* natural to merely natural; a process which is called *naturalization.* Eventually, it was decided this same process would happen in relation to everything scientists ever set out to investigate, be it Life, minds, psychic experiences of various kinds, precognition, telepathy, even near death and out of body experiences.

'All things would be explained in *purely 'natural,'* that is in purely *material* terms and there would be no need to refer to any cause beyond physical nature to explain anything. Unguided, unintelligent causes would explain all. There would be no need, anymore, to refer to Aristotle's concepts of shaping ideas and purposes or to what is today referred to as intelligent design or ID.'

'This, though, led to << THE GREAT ERROR >>. Because to explain what things are *made of* and *how they work,* clever and practically useful as it is, does not, necessarily, tell us how they arose in the first place. Did they have intelligent or mindless causes, random or meaningful causes?

'Some might think this does not matter because, using MM/MN, our sciences have achieved so much in the last couple of centuries. But science is, also, about finding out what is *true*. The latin word, *'SCIRE,'* from which our word science comes, means *to know*. It does not mean materialism. So, is it true that all things are, fundamentally, coincidences and 'unintelligently designed'? Or is there more than luck at work in living things?

'The problem with overusing MM and MN in science today, Ali, is that they are question begging. They assume that nature can be *fully* explained in purely materialistic terms, as Darwin sought to do, when this may not be true.

'What if there are, after all, subtle, intelligent causal factors beyond nature's more obvious physical laws? If so, a purely physicalistic description of the living world will never be complete.

'The problem is that, over the last few centuries, science has concluded that not only is it a useful starting point, as Sir Francis Bacon suggested, to assume that many natural phenomena can be explained in purely materialistic terms, this once useful way of thinking has gone much further than Bacon and his contemporaries intended.

It is now more or less a dogma of our sciences that there are *no* phenomena which do not have purely physical explanations and that, probably, there are no non-physical components to reality at all, a view which leads from methodological materialism or MM to philosophical materialism or PM.

'In recent times the materialist views of thinkers like Darwin have been more fashionable than those of non-materialists like Plato (theory of Ideas, Platonic Forms), Galileo (helio-centricity), Newton (laws of motion, gravity, *Principia Mathematica*), Faraday (electromagnetism, electrochemistry), Maxwell (mathematical-physics, electromagnetic-radiation), Planck (quantum physics) and Lodge (radio, spark plugs).

'As Francis Crick, one of the pioneers of DNA and, himself, a famous materialist put it, "The ultimate aim of the modern movement in biology is to explain all biology in terms of physics and chemistry." [72] This explains why it has become so unfashionable and reputationally risky for scientists to investigate paranormal phenomena like telepathy, spiritual healing and out of

---

[72] Crick F. 2004. *Of Molecules and Men*. Prometheus, p. 10.

body experiences and why those, like Meyer, Behe and Dembski, who put forward various arguments for intelligent design tend to be ignored.

'Right now, materialist thinking dominates and if anything occurs which seems to defeat a purely physicalistic understanding, either such an explanation will be forced onto it or it will be treated as a delusion or hoax.

'This is why MM and MN, when taken too far, can be damaging to science. They close down other possible lines of inquiry and they severely limit the further advancement of our knowledge.

'Unfortunately, under these restrictive conventions, if scientists even begin to wonder whether something may *not* be susceptible to a wholly physicalistic explanation, they are encouraged either to talk themselves out of such heretical thoughts or to maintain that they wouldn't be clever enough to detect any evidence for the supernatural even if it's there! A defeatist attitude surely.

'Take psychic abilities and supernatural experiences of various kinds. Conventional physical scientists have never been able to find any purely physical explanations for such things. So, the tendency is either (a) to ignore them, (b) deny them [73] or (c) to accuse those colleagues open-minded enough to research them of being deluded, susceptible to hoaxes or fraudulent.[74]

'Yet, if something *does* have non-physical causes, we will not be able to explain it in purely 'natural' terms. If confronted with a never-before-seen mechanism, like a Car, desert islanders could, like Darwin, take a purely naturalistic approach and hypothesize that the unfamiliar mechanism must have arisen due to wholly 'natural' processes. They could take it apart, tumble its pieces around in a container, and see if it would 'naturally' re-assemble into its originally very purposefully arranged sequences of parts due to the *natural* laws of physics in combination with the *natural* laws of random tumbling.

'Would this work? Such experiments have been tried with living things, like bacteria, random swirling of the disassembled bacteria parts carried out, electrical sparks fired, and no new life forms have 'naturally' emerged.

'However, for faith-based materialists, closed to all evidence for intelligent causation, and to the supernatural in general, this makes no difference. For them, only materialistic theories of Life will do. This is why they will futilely continue to try to explain *all* the data of reality, *even their own Bright Minds and Immaterial Ideas,* in terms of purely mindless and purposeless material forces. Mindless atoms of Oxygen, Carbon and Hydrogen randomly bumped into each other, they will continue to say, and, eventually, became Michael

---

[73] Did a neurosurgeon go to heaven? Why a Near-Death Experience Isn't Proof of Heaven. M. Shermer. *Scientific American.* April 13, 2013.

[74] Carter, C. 2012. *Science and Psychic Phenomena: The Fall of the House of Skeptics.* Inner Traditions.

Angelo, Isaac Newton and Albert Einstein – all for no reason or point at all. This irrational scheme doesn't work and it's provably impossible, but so it is.'

'What about the religious? Where do they stand on all this?' Asked Alisha.

'Well, some follow Darwin, minus his materialism, and some not. But, either way, for religious people, things can be complex. On the one hand, they believe in the supernatural. On the other, many of them believe that science cannot tell us anything about the subtle, *non-physical* realms of reality.'

'Why not?'

"Because science," they say, "can only concern itself with the physical." As to supernatural matters, only religion, they feel, can give us any answers. Answers which we either accept, on *faith,* or reject.

'Not all those with religious beliefs see it this way. All the well known modern advocates of intelligent design, while they agree that science cannot confirm their own or anyone else's particular religious convictions, believe that today's scientists are perfectly capable of working out whether or not there is *more* to life than the purely physical and whether or not certain features of nature are more likely to have intelligent causes, than random ones.

'Contemporary intelligent design thinkers, like Behe, Meyer and Dembski, simply do their best to demonstrate, using modern methods, that only *intelligence,* not mere chance, could give rise to living things.

'Until only a century or so ago, their ideas would not have been controversial. Pretty much everyone accepted that life really *was* a miracle, not merely the vanishingly unlikely product of chance. The present materialist faith that mere luck can assemble anything remotely clever would have been considered highly unreasonable.

'But, currently, the mainstream science and media communities tend either to misunderstand modern ID thinking or to ridicule it. Look up 'Intelligent Design' in Wikipedia, for example, and [at time of writing] you'll find a very unfair and misrepresentative view of ID. More on that later.'

'So what's your conclusion?' Mused Alisha.

'That science does not *have* to restrict itself to investigating only the physical dimensions of reality. That although, as Sir Francis Bacon intuited, many phenomena can be explained quite well in purely physical terms, and it can be useful to assume, at the start of our inquiries, that it may be possible to explain many things by such means, this approach may not take us *all* the way.

'For example, once we have exhausted all reasonable avenues for finding a purely physical explanation for something, why not allow that some underlying or overarching intelligence(s) may be involved, even if it cannot be directly seen or easily explained?

'Unlike those who conflate science with materialism, why not stay open to the classical view that there may be more to Life than the purely material. If *Mind and Soul* are ontologically real, while they can be known in this dimension – looking at children or animals at play is one way, doing some thinking or being creative are other ways – it may not be possible to reduce them to matter alone as, in recent times, our sciences have sought to do.

'The same for all living things, for the modern evidences for out of body experiences and for afterlife communications from those passed on. Why should our sciences not investigate these kinds of things, using modern, non-religious, methods? Materialism, by contrast, unduly restricts our sciences.

It says, "What's the point of trying to investigate the non-physical dimensions of reality? How do we know if they even exist? We can't physically see them, can we?"

'This, though, is not only defeatist, it's question begging. What if there are non-physical dimensions which, by definition, cannot be seen physically but which can be detected and investigated in various alternative ways? What if materialism isn't true? Are we, then, to train ourselves, as scientists are currently taught to do, to disregard all evidence for intelligence and design in living things, and the evidence for many other kinds of *super* natural phenomena as well, just because it disagrees with the materialist belief system?

'Are we to dismiss every single psychic, spiritual, religious and mystical experience anyone has ever had just to try to make materialism seem true? Are we to ignore every meaningful coincidence, telepathic, clairvoyant or out of body experience just because materialism is inadequate to accommodate that kind of data?

'Materialism is a belief about the most basic nature of reality. So, given this, it needs to be tested. However, if it *is* tested, and it is found wanting, it may be that there are other, better ways of making sense of existence.

'A key test, by the way, is that materialism is self-refuting. Why? Because it is a belief system which, like all beliefs, consists in ideas. Ideas, though, are not physical things, they are abstractions. Therefore, the very phrases, 'materialist philosophy,' 'materialist ideas' and 'materialist beliefs' refute materialism. In any case, modern physics shows us that, in reality, there *is* no matter. There are just multiple forms of *intelligently in-form-ed energy-space.*

'Darwin's contradictory idea of a progressive, yet supposedly aimless evolution from mindless particles to *bright,* purposeful people, is, in any case, illogical. Surely this self-contradictory scheme should be backed up by some extraordinarily good evidence, as Sagan advised, before it is accepted?

'Materialism is also odd because the idea that it is scientific to hold that mere chance can create complex mechanisms is so clearly counterfactual. All our data tells us this doesn't work. The idea that our bright minds randomly formed from the *mindless rocks and lifeless liquids* of our planet by pure luck is illogical, and it's not backed up by any evidence. Yet, it is a very mainstream view today. A view which is constantly presented in our schools and universities and to the wider public as 'scientific' knowledge, as 'real' knowledge, as trustworthy, respectable knowledge.

'To the world's philosophical idealists, though, materialism is self-refuting and it is very easy to falsify. But materialists, from Darwin to Dawkins, justify their faith in pure pointless chance, as a supreme creative principle, because Life's subtle spiritual sources, if real, are not obvious. Of course! We know this. If it were otherwise there'd be no debate at all.

'Materialists also argue that, quite apart from the physical invisibility of life's deeper sources, if there are any, they cannot bring themselves to believe in anything beyond the physical, in anything spiritual, in any positive, final, creative energy or source, which is not only incredibly creative and intelligent but also loving and good, due to life's many hardships.

'Darwin suffered the early loss of several of his children and this deeply affected him. So it's easy to understand why he, like others going through difficult life experiences, can tend to a gloomy, glass half-empty view of reality. He found it hard to believe that there could be anything beyond this often difficult and challenging world. But this, by itself, did not make his views true.

'By contrast, the world's spiritualists argue that the materialist conception of a final Uncaused Cause – whether we label it as God, the Tao, Buddha Nature or the Great Spirit – of tremendous creative potentials, *which just mysteriously Is,* is far too simplistic.

'No serious thinker has ever claimed that the subtle causes of reality manifest, quite literally, as a cosmically scaled Father Christmas type figure, sitting on some cosmic cloud designing Life Forms, or that existence is not both challenging and deeply mysterious.

'They *do* say, though, that there are various ways to make at least some sense of both natural and human evils and that we do not always have to physically see something before we can intuit or logically infer its existence.

'There are plenty of ways to test for meaningful design and meaningful patterns versus pure chance when studying the origins of living things and plenty of ways, too, to conduct rigorous, repeatable research into various supernatural phenomena. We'll consider some of the modern, non-religious research into these fascinating topics a little later.'

We walked on through the lovely valley.

10 Darwin's Mistake: Micro vs Macro

'While Sir Francis Bacon's suggestion to scientists, back in the 1500s, to focus more persistently than ever before on *matter and mechanism,* to really get down to working out what things were 'made of' and 'how they worked,' was a very useful idea for its time, an idea which helped modern science to really take off, people eventually forgot that working out Aristotle's first two causes properly did not tell us anything about his second two causes, shaping ideas and purposes. Sir Francis Bacon's approach did not, necessarily, explain how various natural things, especially living things, had arisen in the first place.

'However, taking an overly materialistic line in science has now become a block to many perfectly reasonable lines of inquiry. For example, desert islanders might be able to work out what a never-before-seen Car was 'made of' and how it 'worked,' and our scientists the same for the Living Cell, for eyes and ears, for blood and brain, but this does not amount to an explanation of their *origins.* Some of the islanders might argue that the Car must have 'fallen together' by pure chance, by purely 'natural' means. Such an 'explanation' would, of course, be mistaken. Others, despite the physical invisibility of the Car's original designers, might argue for intelligence.

'The same applies to living things. Too many people now believe that our science's impressive understandings of 'made of' (amino acids, polypeptide chains, intracellular protein machines) and 'how it works' (DNA CODING, enzyme functioning, digestion and respiration etc) amount to full explanations of how things arose in the first place when, clearly, they do not.

'Cleverly working out what cells or hearts or lungs or brains are 'made of' and how they 'work' does *not* tell us how they first arose. A question to which there are only two possible answers. Either they arose by pure chance, as our current science, taking its lead from Darwin, believes, or they arose due to intelligent causes. The intellectual waters are further muddied by the confusing conflation of micro-evolution with Darwin's theory of macro-evolution.'

'What's the difference?' Asked Alisha, looking intrigued.

'Micro evolution describes relatively minor variations *within* species or genera. Macro evolution is Darwin's far more dramatic idea that all species today living descended from just one or two ancestor microbes. This is the idea that random genetic copying errors (mutations) gradually turned single microbial cells into *multi-cellular* organisms like worms and fish. Further mutations gradually turned water-breathing fish into air-breathing amphibians, dinosaurs into birds, with *completely* different lungs again, burly, land-living bears or some other mammal into deep-sea diving whales, shrews into upside down hanging, fast flying, sonar equipped bats and so on. Only .., there's no evidence for any such processes.'

'So why did Darwin believe such transformations might be possible?'

'Well, he partly based his theory on his observations that, by careful, goal directed breeding, thoughtful, intelligent, people could bring about quite considerable changes to the form of an already existing species.'

'Like wolves to dogs or pigeon or cattle breeding?' Asked Ali.

'Yes. But Darwin made no allowance for the fact that clever human selectors, even with very clear aims in view, have only ever been able to change existing species up to *certain genetic limits* which they have never been able to cross without damaging or destroying the animals in question. Such *within* species changes, wolves to dogs or finch beak variations, are examples of micro-evolution. But there is no evidence that microevolution, natural or artificial, has ever led to macro-evolution, to an entirely *new* type of animal.

'Unfortunately, Darwin ignored this key point and made an unjustified extrapolation from micro to macro evolution. Unjustified because the fossils have never confirmed his idea. The "inconceivably great" numbers of transitional species his theory requires are never to be found and it disregards the strict limits to the changes which can be wrought by artificial breeding.

'It has never been possible to artificially evolve a species beyond the limit set by its particular gene pool without damaging it or destroying it. Wolves to Chihuahuas? Yes. Wolves to a new species? No. Single cell microbes to multi-celled worms, trilobites or fish? It's implausible and, more importantly, there is *no* fossil or other evidence for it. Land running shrews to fast flying, echo-locating bats? No evidence. Bears to whales? No evidence.

'Darwin's idea was that if an existing creature, like a Wolf, gave birth to a different version of itself, like a Chihuahua, it might survive and become a new species. The evidence, though, is that natural selection tends to *confirm* the species type, except in respect of minor variations, not to deviate it to any great degree, so only the best adapted members of an existing species survive.

'Sickly or maladapted variations of the original kind, like Chihuahuas, will not survive in the wild and healthy variations will be even more Wolflike than before, faster and cleverer. Yet, as Michael Flannery writes in his article, *'Was Darwin a Scholar or a Pitchman (skilled salesman)?,'*

'For Darwin, the fact that man could breed a fancy pigeon or an especially fast race horse or a unique dog indicated evolution "in action."

But, as Wallace pointed out to him, when left in the wild, these fancy breeds [like Chihuahuas] either perish or revert to their original type. Besides, domestic breeding of animals requires the very thing Darwin sought to avoid – *careful thought and pre-selection.* In effect, it requires a breeding plan and design. This is clearly *not* random and purposeless, wholly natural causes operating to produce speciation. Darwin never saw that logical flaw in his own theory, yet it was obvious to naturalist [Alfred Russell] Wallace, zoologist Pierre Grassé, historian Jacques Barzun, and many others.' [75]

'That's a serious charge against Darwin, usually so revered.' Mused Ali.

'Yes, it is. But Darwin found it hard to see or to accept some of the most obvious flaws in his own theory. And, crucially, the fossils have never yielded any evidence for the kind of vast and 'indefinite departure' [76] from an original life-form which, he believed, might turn single-cell microbes into worms, spiders or fish, or, ultimately, into you and me. In fact, as biologist James Shapiro says, cells go to very purposeful and clever lengths to,

'protect themselves against precisely the kinds of accidental genetic change that, according to conventional [evolution] theory, are the sources of evolutionary variability [from an earlier form to give rise, as Darwin maintained, to a wholly new creature. No. Cells] ... devote large resources to *suppressing* random genetic variation.'[77]

'Not only this, but all human attempts to breed or to evolve any species beyond a certain point have never resulted in a new species emerging but only in existing ones being damaged. Dogs and fruit flies are famous examples.

'It is true that in the wild, when conditions vary, relatively minor changes can take place *within* existing species due to the potential for adaptive variability built into their existing gene pools.

'If the bark on trees becomes darker, as happened during the industrial revolution, the darker peppered moths, already in existence, do better than the lighter ones which get eaten more often than the darker ones. When the pollution ends, and the tree barks become lighter, the fairer moths, which have never gone away, begin to predominate again.

'These kinds of minor, reversible, changes do not depend on random genetic mutations but take advantage of the moths' existing genes, taken from

---

[75] Michael Flannery. 'Was Darwin a Scholar or a Pitchman?' *Evolution News* (October 20, 2015). Emphasis added.

[76] Bethell, T. 2016. *Darwin's House of Cards.* Discovery Institute.

[77] Quoted by Lennox, J. 2009. *God's Undertaker.* Lion; p. 143, emphasis added.

a gene pool which includes possibilities for darker and lighter examples to hatch, the surviving proportions of which change as conditions change. But the proposal that this kind of minor change proves that an ancient microbe could either (i) arise by chance or (ii) gradually morph into a very different trilobite, worm, fish or spider, also by chance, is not backed up by the data.

'But peppered moths and finch beak variations aside,' asked Ali, 'are not the equine fossils and the ancient bird, archaeopteryx, evidence for Darwin's theory?

'Well, regarding horse fossils, in *The Naked Emperor: Darwinism Exposed,* Antony Latham sees this as microevolution. He writes, 'We do not see in the horse series any major change in anatomy, just loss of digits, increased size and altered tooth shape. This hardly constitutes evidence for Darwinian evolution, except on the micro scale.'[78]

'Jonathan Wells observes that Henry Gee of *Nature* magazine, while himself a believer in Darwin, 'candidly admits that we can't infer descent with modification [for horses or any other species] from fossils.'[79]

"No fossil is buried with its birth certificate," he wrote in 1999. ... To take a line of fossils [such as those relating to equines] and claim that they represent a lineage is not a scientific hypothesis that can be tested but an assertion that carries the same validity as a bedtime story – amusing, perhaps even instructive, but not scientific."[80]

'His point is that various fossils, which can be set out to appear like a succession, do not tell us that the later, similar species directly emerged from the earlier forms. The primary issue, though, is not succession but *cause*. We create and evolve things in successions all the time. All modern Fords are successors of Model T. But every Ford since Model T is an intelligently designed and separately created model or species in its own right.

'Are you denying, then, that species can vary?' Asked Ali.

'No. There are minor, backwards and forwards variations in micro-evolution: finch beaks, peppered moths, wolves to dogs. What is not yet widely admitted, though, nor passed on to young people or the general public, is that there is no evidence for Darwin's key idea that one kind of animal can gradually turn into a wholly different type.

---

[78] Latham, A. *The Naked Emperor: Darwinism Exposed.* Janus, p. 88. Text in square brackets added.

[79] Wells, J. 2006. *The Politically Incorrect Guide to Darwinism and Intelligent Design.* Regnery Publishing.

[80] Gee, H. 1999. *In Search of Deep Time: Beyond the Fossil Record to a New History of Life.* The Free Press; pp. 32, 113–17.

'NOWHERE has any fossilized species been found which is on the way to becoming a spider [or snail or fish or eagle or bear or whale]. It's spiders or something else! All creatures, not just spiders, are well structured wholes.'[81]

As to Archaeopteryx, in *Icons of Evolution: Science or Myth?*, Jonathan Wells explains,

'. . . there are too many structural differences between Archaeopteryx and modern birds for the latter to be descendants of the former. In 1985 University of Kansas paleontologist Larry Martin wrote: "Archaeopteryx is not ancestral of any group of modern birds." Instead it is "the earliest member of a totally extinct group of birds."'[82]

'So, if archaeopteryx is not an ancestor of modern birds, is he a descendent of older dinosaurs?'

'Good question. In *Icons*, Wells discusses the idea that archaeopteryx is descended from a group of dinosaurs the only fossil evidence for which comes in a stratigraphic period long after archaeopteryx was extinct.'

'So how could later dinosaurs be ancestors to a much earlier animal which was already extinct when they were around?' Asked Ali.

'Well, Darwinian proponents of this idea say it's because the fossil records are incomplete. But Wells is skeptical. He writes,

'The claim that birds are dinosaurs strikes most people – including many biologists – as rather strange. Although it follows from cladistic theory, it defies common sense. Birds and dinosaurs may be similar in some respects, but they are also very different.'[83] [For example, birds have *unique lungs*, different to those of all other vertebrates, including dinosaurs.]

'Darwin believed that species could vary *indefinitely*. Tom Bethel, in *Darwin's House of Cards*, calls this "DARWIN'S MISTAKE." He writes,

'We all know that small differences are observed between the generations. ... ***Darwin's mistake*** was to assume that these differences somehow accumulate over the millennia, so that one species eventually transforms itself into another.

'***Without evidence***, Darwin's supporters today still [believe this]. ... But such a transformation has never been observed [either in history or in the

[81] Davidson, J. 1992. *Natural Creation or Natural Selection – A Complete New Theory of Evolution. Element;* p. 13–14, emphasis and text in square brackets added.

[82] Wells, J. 2000. *Icons of Evolution.* Regnery, p. 116.

[83] Ibid.

world's fossil records]. *No species [fossil or living] has ever been seen to evolve into another.'* [84]

'What the fossils do show, Ali, are reversible, back and forth changes *within* species and the mysteriously sudden and, at times, revolutionary appearances of new forms, *without* plausible biological ancestors. Existing species, aside from minor changes, *stay* as they are. Wolves remain wolves, fruit flies remain fruit flies and bacteria remain bacteria. Darwin wondered if an existing species might gradually morph into something wholly other. A process he called natural selection. But this is another example of his muddled thinking.

'Why muddled?'

'Because the aimless nature of materialist thinking cannot *'select'* anything. How can it? Remember, it has no goal or aim in mind. It has no mind! Yet, artificial breeding, which *does* involve selection in the true sense, has never managed to create a new species, let alone a new genus or family.

'Darwin was well aware, based on the best breeders' real-world experience, not theory, that *no species* has ever been bred to vary beyond a certain point without damage. He, though, did not wish to agree to the implications of this.'

'Why not?' Wondered Alisha.

'Because it tended to falsify his theory. But it does explain why he could not *'point to a single case of the unlimited variation that his theory required.'* [85]

'As science writer Richard Milton notes,

'Darwin's theory of evolution relies on the idea of random genetic mutation... altering species... But all experimental evidence indicates that the extent of genetic change wrought by natural selection is quite limited. . . . [it is also an] embarrassing fact that . . . *no transitional species showing evolution in progress between two stages has ever been found* . . .

'[The] theory of evolution has become *an act of faith* rather than a functioning science. . . . Not until the scientific method is [properly] applied to Darwinism will it be exposed, and only then will the right questions be asked about the mystery of life on earth.' [86]

We walked on.

[84] Bethell, T. 2016. *Darwin's House of Cards*. Discovery Institute, emphasis and text in square brackets added.

[85] Ibid.

[86] Milton, R. 2000. *Shattering the Myths of Darwinism*. Inner Traditions Bear and Company; back cover, text in brackets added.

11 Genetic Copying Errors

'If we remain open, Ali, to the classical idea that reality includes subtle, non-physical aspects, we can allow for the possibility of intelligent causation. We can also try to see, using modern methods, if we can find any scientific evidence for design and intelligence at work in living things.

'But, currently, the science method conventions known as methodological materialism and methodological naturalism, MM/MN, oblige scientists to attempt to explain *all* aspects of Life, *even Mind,* as though they are purely materialistic phenomena, even if they may not be, and even if there are key non-physical components to reality. Thus, Darwin speculated,

> '*Might* a complex microbe arise, by unguided chance, from *non-living chemicals,* in some ancient, lightning-struck pond? *Might* it randomly vary, until it turned into an even more complex worm, trilobite or fish? *Might* a fish randomly vary to become an air-breathing amphibian?
>
> '*Might* that give birth to dinosaurs? *Might* the dinosaurs vary to become birds with wings and feathers, and *altogether different* lungs again? *Might* other dinosaurs randomly vary to become mammals? *Might* some of these turn into apelike creatures and, eventually, people, *all by chance?'*

'And, pursuing these kinds of speculations, where Life is thought to be just another random effect like rain or snow, which can be explained in purely physical terms, Darwin hoped the fossils would supply an "inconceivably great" (his words) number of transitional species to document his ideas.

'Unfortunately for his views, the fossils have never obliged. The millions of linking species and their ancestors which ought to be found at the hypothetical branching points on his hypothetical tree of life, are never to be found.

> 'Now, after over 120 years of the most extensive and painstaking geological exploration of every continent and ocean bottom, the picture is infinitely more vivid and complete than it was in 1859. Formations have been discovered containing hundreds of billions of fossils and our museums now are filled with over 100 million fossils of 250,000 different species. ... [So] What is the picture that the fossils have given us? Do they reveal a continuous progression connecting all organisms to a common ancestor? With every geological formation explored and every fossil classified it has become apparent that

these, *the only direct scientific evidences relating to the history of life, still do not provide any [of the] evidence* for which Darwin so fervently longed. [87]

'So there really is no fossil evidence for one species gradually turning into another, over and over again, as Darwin predicted?' Commented Alisha.

'No. If it were otherwise, Ali, I would not argue with Darwin. Because I'm not a one-off creationist, who believes creation can happen only *once*. Why should subtle creative forces, to which people assign various religious names, not *evolve* things, including Living things? The difficulty is not with all notions of evolution but with Darwin's ideas. Firstly, due to the lack of evidence and, secondly, because, as a materialist, he *had* to rely on pure luck for all his evolutionary creativity. But, as we've seen, this doesn't work, it's impossible.

'This is why we need to concede that natural selection, with no aims in mind – for it has no mind – would be inadequate to *create* a highly complex common ancestor microbe, in the first place, let alone to *evolve* it into a DNA CODED worm, trilobite or spider and, later, creatures with the minds, habits and instincts of bears, elephants and whales, let alone human beings.

'It is also clear, by now, that a species or, at best, genus can only vary *within* but *not beyond* the limits set by its original gene pool. It is such inbuilt, but limited, genetic variability which enabled people to derive today's diverse dog breeds from their original wolf stocks, but all dogs are still, in essence, wolves, *Canis Lupus*. None of them is a new species. It's such limited variability which accounts for Darwinism's famous finch beak and peppered moth variations.

'But such minor micro-evolutionary variations, which are uncontroversial, cannot explain how Darwin's hypothetical common ancestor microbe might, as he believed, dramatically turn into altogether different Snails, Fish, Whales, Flowers, Trees, Monkeys or People. Further, modern Darwinism, which attributes such extraordinary creative powers to random genetic mutations, doesn't explain where the original, healthy, non-mutated genes, and their highly complex DNA CODED instructions, came from in the first place.

'It just tells us they 'must have evolved' and it hopes we won't notice the intellectual dissonance. Because, even archaic bacteria, we now realize, are vastly more complex than Darwin and his steam era contemporaries imagined. It is clear they could *never* arise due to some purely chance based process.

'Lastly, intelligently guided wolf to spaniel or chihuahua micro-evolution takes advantage of the wolf's existing gene pool but it *doesn't* go beyond it. Spaniels and chihuahuas are, essentially, still wolves, not a new species, new genus or family let alone a new order, class or phylum. For an entirely new

---

[87] Sunderland, L. 1988. *Darwin's Enigma: Ebbing the Tide of Naturalism.* Emphasis added.

Life Form to arise in line with Darwin's ideas, for a *single* cell microbe to turn into a multi-celled worm, trilobite, snail, caterpillar or spider, completely *new* genetic information, far beyond the original cell's gene pool, would be needed.

'Darwin, though, had no idea how the random variations his macro-evolutionary theory called for might arise and, eventually, via millions of hypothetical biochemical changes, and a vast number of transitional forms, turn an ancient microbe into a snail, fish, stegosaur, lion, eagle or whale.

'Subsequent to Darwin, Gregor Mendel's pioneering work on genes was rediscovered. It showed how variations within species' existing gene pools arose and how it was this natural shuffling of *existing* genes which had been artificially selected by human breeders in their pursuit of new and better varieties, different kinds of dog, pigeon or tulip for example.

'It was this natural gene shuffling of existing genes which favored the increase of those species members with the mix of genes best suited to the prevailing conditions and it was this same process which accounted for the kind of minor changes seen in peppered moth and finch beak variations.

'Mendel's discovery could not explain, though, how a single-cell microbe or a multi-celled fish might gradually change into an entirely *other* kind of creature, microbe-to-fish, fish-to-amphibian, amphibian-to-dinosaur, dinosaur-to-bird-or-mammal, mammal-to-ape, ape-to-human.

'Why not?'

'Because Mendel had simply worked out how healthy, *correctly copied* genes, from within a species' *existing* gene pool, accounted for minor *within species* variations. He had not, though, found any evidence that such shuffling of existing, correctly copied genes could cause a species to be gradually changed, – either by mindless, unguided nature, or by careful, goal-oriented, human breeders, – into an entirely new species, genus, family, class, order or phylum.

'Some scientists then wondered whether random genetic mutations might not take place, create *completely new* genetic information and lead, gradually, via many transitional versions of the original species, to entirely new plants or animals. An idea which came to be called neo-Darwinism.'

'Why 'neo' Darwinism?'

'Because Darwin had no knowledge of genes. But, apart from the never to be seen fossil evidence for the transitional species called for by Darwin, random genetic mutations seem to be *relatively rare genetic copying mistakes.*

'Genetic copying errors which are generally harmful, often lethal, and no more do they provide useful new constructional information than a poorly copied circuit board in a computer would be a potentially useful new feature.

'In *God's Undertaker: Has Science Buried God?* John Lennox notes that 'The incredibly precise duplication of DNA is not accomplished by the DNA alone: it depends on the presence of the living cell.' He quotes James Shapiro,

'It has been a surprise to learn [just] how thoroughly cells protect themselves against precisely the kinds of accidental genetic change that, according to conventional [neo-Darwinian] theory, are the [very] sources of evolutionary variability. By virtue of their ***proofreading and repair systems***, living cells are not the passive victims of the random forces of chemistry and physics. They ***devote large resources to suppressing random genetic variation*** and have the capacity to set the level of background localized mutability ***by adjusting the activity of their repair systems***.'[88]

'So it's unlikely that random genetic mutations could drive the kind of macro-evolutionary process which Darwin had in mind?' Mused Ali.

'Correct. Yet, currently, any scientists wondering whether these complex error correction processes point, not to 'luck' but to fascinating intelligence at work in living things, risk being told they are being 'fanciful' and 'unscientific.' No, they are required to keep materialist discipline and to assume, in defiance of common sense, that somehow or other these clever proof-reading systems arose by mere 'chance' and that they are 'purposeless.'

'Remember, ***according to modern scientific convention,*** dictated by methodological materialism and methodological naturalism, MM/MN, ***nothing in Nature has any purpose or teleology***. Which means that all Life is, supposedly, completely accidental. Pure, stupid luck can, and does, do it all.

'Yet random genetic mutations, marketed so widely and glibly as the engine of an amazing evolution, culminating in the intellectual and creative capacities of a Plato, Galileo or Picasso, it has been amply demonstrated, cannot build the new and any minor advantages, which they can confer, come at a cost.

'In *The Edge of Evolution: The Search for the Limits of Darwinism,*[89] biochemist Michael Behe explains that random mutations are ***not constructive***. He cites, for example, the dangerous sickle cell mutation – which has the one benefit that it can confer immunity to malaria – to show how *uncreative* random genetic mutations are. They are evidence for the *loss* of genetic information, not for construction.

'Although Darwin's theory can explain minor evolutionary change, random genetic mutations could not drive the kind of evolution he had in mind, let alone account for the basic machinery of Life. Behe adds that there is *no*

---

[88] Lennox, J. 2009. *God's Undertaker.* Lion; p. 143, emphasis added.

[89] Behe, M. 2008. *The Edge of Evolution.* Free Press

evidence that the natural barriers between the species, up to the classification level of families, could ever be breached by *any* kind of random process.

'In fact, despite their focused, persistent, efforts, clever human selectors have never been able to breach even the species barrier – chihuahuas are still essentially wolves, pigeons are still pigeons and fruit flies are still fruit flies.

'Dr. Durston, a scientist and philosopher writing in *Evolution News* notes,

> 'There is mounting evidence that most, if not all the key predictions of the neo-Darwinian theory of macroevolution are being falsified by advances in science. ... Ask computer programmers what effect ***ongoing random changes*** in the CODE would have on the integrity of a program, and they will universally agree that it ***degrades the software***. ... deleterious mutations [do not create biological life, they destroy it].' [90]

'As Dr Durston points out, random genetic mutations are generally harmful, not beneficial, and, rather than causing a genetically deformed creature to be turned into a marvelous new Life Form, would lead to it being eaten!

'Even antibiotic resistance in bacteria is caused by minor genetic typos which, as Dr Behe explains, reverse when the antibiotic is no longer used.

'Ironically, as Dr Behe points out in his book, *Darwin Devolves,* Darwin did not actually *show* that living mechanisms, be they single cells, multi-cellular creatures or amazing organs like hearts and brains, livers and lungs, eyes and ears, *really could* be built by mindless natural selection acting on random chemical variations – for no reason or purpose at all! No, Darwin simply argued, ***without any convincing evidence,*** that this 'might' be possible.

'Why though?'

'Because, as a philosophical materialist, Ali, he *had* to. Materialists, like Darwin and, more recently, Richard Dawkins, *have* to try to fill all their knowledge gaps with endless appeals to mindless chemical coincidences to 'unintelligently' and 'purposelessly' build everything, from the bee's sting to the eagle's wing, from the mammalian eye to the human brain.

'For Darwin, living things might *look* designed and full of purpose, but this was an illusion. Behe quotes Darwin who wrote, "There seems to be no more design in the variability of organic beings and in the action of natural selection, than in the course which the wind blows." Pure chance could do it all, he felt.

'As Dr Behe also points out, while Sir Francis Bacon suggested, at the dawn of the modern age, that scientists need not worry about the ultimate ideas and purposes of things, be they Atoms, Rocks or Horses, in order to do good

---

[90] Durston, K. 9.7.15 An Essential Prediction of Darwinian Theory Is Falsified by Information Degradation *Evolution News.* Emphasis added.

quality, practical, everyday science on them, to deny that *anything* in nature, even *living* nature, has any *teleology* or purpose is clearly counterfactual.

'Because, as he explains, while we may not know what a Horse is 'for,' we do know that its heart and its lungs, its mouth and its legs *do* all have purposes which make the *fascinating, living, mystery we call the Horse,* possible.

'In *Darwin Devolves,*[91] Behe explains that random genetic copying errors (mutations) *will* be naturally selected, and spread through a population, when they confer an immediate advantage on a species. *But, this always comes at the cost of the loss of genetic information.* He argues that random mutations cannot build complex new structures such as eyes or brains. Why not? Because, while genetic mutations sometimes provide a marginal advantage to an *existing* organism, essentially, *they degrade and devolve* – hence the title of his book.

'This is why the modern Darwinian hypothesis that a dinosaur's foreleg could turn into a bird's wing due to a long series of rare genetic copying errors – which all cells actively protect against – is unrealistic. *It can no longer be considered good science.* Darwin's far-too-slow-to-die hypothesis persists not because of its scientific merits, which are, by now, very low, but for reasons of philosophy, psychology, professional pride and unexamined belief.

'Not only this, but birds' feathers, exquisitely engineered in their own right, would need to arise, due to yet more random mutations. And, at the same time as this imaginary wing and feather creating process was taking place, the 'unintelligently designed' prototype bird of Darwinian thinking would have to be developing the *unique* avian lungs, different from those of all other vertebrates, without dying more or less instantly in the process.

'Darwin believed that creatures could vary indefinitely. James Perloff, author of *Tornado in a Junkyard,* a title referring to Sir Frederick Hoyle's well known comment that it is as likely that Life could arise by mere luck as it is that a hurricane blowing through a junk yard could assemble a jumbo jet, writes at his website,

"According to Darwinism, single cells eventually evolved into [multi-celled] invertebrates.., then successively into fish, amphibians, reptiles, and finally mammals. Darwin said this occurred from creatures adapting to environments. [But Mendel's] discovery of genetics threatened this claim. New organs require new genes."

'However, "Just moving into new environments doesn't give you new genes. This initially stumped Darwinists, but they eventually [thought they had] found a solution. Random mutations — copying mistakes in the genetic code — occur very rarely, but [they] DO alter genetic information. So modern

---

[91] Behe, M. 2019. *Darwin Devolves.* Harper One.

evolutionists said animals gained new genes by chance mutations . . . which they adapted to evolve into higher forms."

'Yet, he continues, "chance mutations are to the genetic code what typos are to a book: they remove information, but do not improve it. In humans, mutations cause sickle cell anemia, cystic fibrosis, hemophilia, Down's syndrome, and thousands of other genetic diseases . . . even the rare "beneficial mutations" evolutionists trumpet – such as bacterial resistance to antibiotics – actually result from functional [genetic] losses," writes Perloff.

'This is why, based on all we now know, by contrast with Darwin's far more limited 19th century knowledge, it no longer makes sense to imagine that avian lungs, wings and feathers, which are marvels of bio-engineering, could arise by mere biochemical chance. And, there is *no* evidence for any such process.

'The feathered wings of Birds and the featherless wings of Bats *appear in the fossil records suddenly and fully formed, in the usual non-evolutionary way*. There is no evidence to suggest they very gradually evolved from wingless dinosaurs in the case of birds or wingless mammals in the case of bats. As Michael Denton explains in *Evolution: A Theory in Crisis:*

> 'It is not easy to see how an impervious reptiles scale could [as Darwinians claim] be converted gradually into an impervious feather without passing through a frayed scale intermediate which would be weak, easily deformed and still quite permeable to air. It is true that basically the feather is indeed a frayed scale – a mass of keratin filaments – but the filaments are not a random tangle but are ordered in an amazingly complex way to achieve the tightly intertwined structure of the feather. Take away the exquisite co-adaptation of the components, take away the co-adaptation of the hooks and barbules, take away the precisely parallel arrangements of the barbs on the shaft and all that is left is a soft pliable structure utterly unsuitable to form the basis of the stiff impervious feather. The stiff impervious property of the feather which makes it so beautiful an adaptation for flight, depends basically on such a highly involved and unique system of co-adapted components that it seems *impossible* that any transitional feather-like structure could possess even to a slight degree the crucial properties. In the words of Barbara Stahl, in *Vertebrate History: Problems in Evolution,* as far as feathers are concerned, "how they arose initially, presumably from reptiles scales, defies analysis."[92]

'Darwin's supporters argue that birds feathers 'must have' arisen from reptile scales because they are made of similar materials. But no one would

---

[92] Ibid, p. 209, emphasis added.

argue that airplanes evolved from cars – by chance – just because they are made of similar materials. Darwin's supporters also maintain that birds' lungs 'must have' evolved from the standard vertebrate lung because birds and dinosaurs share *some* similarities. This is what Michael Denton says about this:

> 'In all other vertebrates [including dinosaurs] the air is drawn into the lungs through a system of branching tubes which finally terminates in tiny air sacs, or alveoli, so that during respiration the air is moved in and out through *the same* passage. In the case of birds, however, the major bronchi break down into tiny tubes which permeate the lung tissue. These so-called para-bronchi eventually join up together again, forming a true circulatory system so that air flows in *one direction* through the lungs. This *unidirectional* flow is maintained during both inspiration and expiration by a complex system of interconnected air sacs in the bird's body which expand and contract in such a way so as to ensure a continuous delivery of air through the para bronchi. ... ***No lung in any other vertebrate species is known which in any way approaches the avian system.*** Moreover, it is identical in all essential details in all birds as diverse as hummingbirds, ostriches and hawks.
>
> Just how such an utterly different [and irreducibly complex] respiratory system could have evolved gradually from the standard vertebrate design is fantastically difficult to envisage, especially bearing in mind that the maintenance of respiratory function is absolutely *vital* to the life of an organism to the extent that the slightest malfunction leads to DEATH WITHIN MINUTES. '[93]

'Although birds and dinosaurs do share some interesting similarities, avian lungs are all-or-nothing. They are *irreducibly* complex. There is no evidence that they could, or did, arise by any kind of random or unguided process.

'Darwin's unbacked claims that this 'might' be so are no longer enough.

'In light of what we now know, Darwin's hopes for his overall theory can no longer be taken seriously. They are no longer good science. They are, though, preventing us from gaining a better understanding of our true origins.

'The evidence of the fossils, genetics and embryology does not support Darwin's core macro-evolutionary idea of one creature gradually turning into another via an "inconceivably great" number of transitional forms. ***Yet scientists remain enjoined to continue to waste their time, trying to find an explanation for a process, Darwinian macroevolution, for which there is no good data?*** Changes over time? Yes. But to call those changes evolution, in the

---

[93] Ibid. p. 210–212. Emphasis and text in square brackets added.

sense Darwin meant, is misleading, both to the general public and to scientists from other fields who rely on their evolutionary colleagues to keep them well informed.

'James Perloff writes,[94] "Darwin claimed life began eons ago from chance chemical processes. ... This might have been plausible in Darwin's day, when cells were considered simple. But no longer. Even a [basic] bacterial cell requires thousands of different proteins... in precise order. Frances Crick, who co-discovered DNA's structure, estimated the odds of getting just ONE protein by chance as one in 10 to the power of 260 — a number beyond imagination."

'Living cells are run by DNA software, which is far more complex than any software we have ever created. Would anyone argue that our software systems could arise by chance? Of course not.

'If, Perloff adds, "bacteria evolved successively into invertebrates, then fish, amphibians, reptiles, and mammals," as Darwinians claim, "there must have been countless "transitional stages,"" but isn't it odd, he notes, that such transitional fossils are *never to be found,* despite there being billions of fossils to look at?

'He continues, "Think about it. For a fish to become a land creature, turning its fins into legs would require new bones, new muscles, new nerves — and while it was adapting to life on land, *a new breathing system*. Since this supposedly happened from chance mutations — which are rare events — innumerable creatures would have to live and die during the intermediate period.

"So WHERE'S THE EVIDENCE for these transitionals?" He asks.

"Not in the living world. Among bacteria, invertebrates, fish, amphibians, reptiles and mammals, there are many thousands of species, but *no intermediate species* between these groups. ...

"Evolutionists try to explain the missing intermediates by saying "they all became extinct." But, *the alleged transitional links are never to be seen.* While the fossils show variations *within* species, they do not evidence change from one animal group to another as Darwin claimed. For example, while billions of invertebrate fossils exist, fossils illustrating their alleged evolution from simple ancestors are missing." [95]

We walked on in the delightful valley, the river quietly flowing by our side.

---

[94] Perloff J 1999, *Tornado in a Junk Yard* https://jamesperloff.com/why-the-creation-evolution-debate-matters/

[95] Ibid, emphasis added.

12 What Is Life, What Does Science Say?

'It is strange, Ali, but another major disappointment of modern biology, is that, despite its many achievements, contemporary science has no compelling ideas as to what Life or Mind truly are. All it can offer is Darwin's theory, which, despite the misleading title of his famous work, *On the Origin of Species*, does not, in fact, tell us how Life originates. Regrettably, our science has abandoned Cartesian dualism and this leads us into today's intellectual mess and dead end.

'Cartesian dualism?' Asked Alisha, looking intrigued.

'It's the classical idea, described by the french philosopher Rene Descartes, that we can think of the body as a kind of machine, but one which is pervaded by *Life Force, Mind and Soul* principles. According to this conception, physical reality manifests through two principles, conceived of, variously, as Spirit and Matter, Soul and Mechanism, Life and Form, Mind and Brain, Subjective, inner Being married, temporarily, to objective, outer body.

'In classical dualistic thinking there are *Life and Soul* principles, the subtle forces which make living things alive: the *animate* versus the inanimate, the *Soul* which pervades its temporary housing. Currently our science is not interested in such ideas. The dismissal of dualism, especially since Darwin, leads to *materialist-monism* where only the physical is considered real and *all* subjective phenomena, like thinking and feeling, are held to be illusions.

'Yet, should we take seriously the statements of those who claim that anything they think or say is nothing more than the pointless firing of purposeless neurons governed by mindless physical laws? As Francis Crick said, "The ultimate aim of the modern movement in biology is to explain all biology in terms of physics and chemistry," that all we are is a "vast assembly of nerve cells and their associated molecules," "nothing but a pack of neurons," as Lewis Carroll's Alice might have put it.[96]

'The problem with this approach, which has our present science culture so much in its grip, is it's question begging. It assumes materialism to be true and to make its worldview seem true, it has to deny even the conclusions of some of its own followers. Thinkers like Wald, Monod and Crick who, although they conceded that Life could *not* by arise by chance, were obliged to deny their own conclusions in order to remain *faithful* to their belief system.

---

[96] Crick F. 1995. *The Astonishing Hypothesis.* Scribner, p. 3, text in square brackets added.

'It is because of this dogmatic ideology that our present science culture cannot think deeper as to what causes Living things to Breathe, the Heart to beat, the Soul to Feel, and our Awarenesses to be Aware. Our present science cannot have any real idea, even, what it is which makes something *Alive!*

'Because, in its view, there are no subtle *Life Forces, Spirit or Soul* aspects to any Living things, no principles like *Chi or Prana* as the Chinese and Indian civilizations know these things.

'But why dismiss any concept of the Spirit or Soul and Life Forces?' Queried Alisha. 'The idea that such forces inform *all* living things seems obvious. Surely anyone can see that when a Plant, Animal or Person dies, a special something has gone. Some vital, animating force is no longer there.'

'Well, it's mostly because someone artificially synthesized urea in 1828 and because Souls can't be seen coming from the spiritual dimensions in telescopes.'

'Ha, ha, very funny . . . But what has *urea* got to do with it?'

'Well, at one time, before our sciences became quite so materialistic in their thinking, it was believed that the existence of the Life Forces was proven by the difference between the *organic* chemicals making up the forms of the plants, animals and people and the *inorganic* ones which make up rocks and stones. So when, in 1828, someone artificially synthesized urea, an organic chemical from inorganic chemicals, the hypothesis collapsed.'

'That's quite a jump, isn't it?' Said Alisha.

'Yes, because it's not just differences in materials which tell us whether something is of intelligent or random origin but its *functionality*. We don't say no intelligence went into the making of a TV because it turns out that the 'organic' (the organized) wires and plastics in it are actually no different, in their essence, from the 'inorganic' (the unorganized) ores and oils in the ground from which the wires and plastics in the TV were derived by *intelligent* forces using *intelligent* methods. Means far less sophisticated than those required to cleverly organize even the most basic of Life forms – all vastly more complicated than anything we make.[97]

'Yet, according to the contemporary scientific view, if you think it through, we are no more truly alive than the inanimate rocks of which our planet is made. *Anima,* after all, means *Life and Soul* but the current 'scientific' view is that we do not have Life Forces or real Minds or Souls. Why? Because, although people seem to forget that the urea didn't synthesize itself, a human *intelligence* did, the fact that it could be artificially synthesized led some to think this showed both Cartesian dualism and vitalism to be false. Since then,

---

[97] Denton, M. 1985. *Evolution a Theory in Crisis.* Burnet Books, p.250.

in the scientific west, to be accused of being a *vitalist,* to believe in the Life Forces or in the immortal Soul, is to be disreputable and 'unscientific.'

'This is why the modern – non-religious – evidence for near death and out of body experiences and for after-life communications from those passed on, even when verifiable,[98] is not taken seriously. Not because the evidence isn't good, but because it conflicts with materialism. Yet the honest truth is that no one has any idea, scientifically speaking, how either *Life or Mind* arise, in the remote past or now. Did you know that there are, literally, *trillions* of self-directed biochemical operations happening in our bodies *this* second and *every* second, 24/7, to keep the whole astonishing thing working?'

'I guess I never thought of it that way.' Alisha said, thoughtfully.

'This is why it's so unreasonable – and so obviously unscientific – to continue to maintain that Life might arise by chance, that it's nothing clever or in any way designed, or that microbes might just randomly morph into Spiders with their extraordinary capacities, or, via numberless imaginary transitional forms – *which are never to be found in the real world's fossil records* – into Lions and Tigers with their Cat consciousnesses, into Wolves with their intriguing Dog natures, full of potential love and friendship for people, and, finally, into Humans with all their extraordinary gifts.

'Remember, there is no evidence *at all* that pure chance can create anything remotely intelligent. Recall, as nobel laureate George Wald said,

> 'When it comes to the origin of life there are only two possibilities: Creation or spontaneous generation. There is no third way. Spontaneous generation was disproved one hundred years ago, but that leads us to only one other conclusion, that of supernatural creation. We cannot accept that on philosophical [ideological] grounds; therefore, we CHOOSE TO BELIEVE THE IMPOSSIBLE: That life arose spontaneously by chance!'[99]

'Unfortunately, this way of thinking, which continues to dominate our sciences, is much more a matter of belief and philosophy than many realize.

'Like so many other believers in pure stupid luck as a supreme creative principle, George Wald didn't *know* that intelligent causality is not possible. It was just that his materialist philosophy did not allow him to consider it.

'For Darwin to wonder whether Life or Mind might be purely *naturalistic* phenomena, not different, in their essence, from other natural phenomena like wind and rain, avalanches or forest fires, or that mindless matter + mindless

---

[98] Schwartz, G.E. Phd. W.L. Simon. 2002. *The Afterlife Experiments: Breakthrough Scientific Evidence of Life After Death.* Atria Books. Fontana, D. 2005. *Is there an Afterlife: A Comprehensive Overview of the Evidence.* O Books.

[99] Quoted from George Wald, 'The Origin of Life', *Scientific American* 191:48 (May 1954).

laws + mindless chances might be enough to explain all, including our own intelligent minds!, was, just about, understandable. Because he knew so much less than we of the truly astonishing complexities of living things. But for us to continue to "to believe the impossible" is, in our day, deeply unreasonable.

'I remember seeing Richard Dawkins, one of the most famous modern proponents of the materialist worldview, get quite cross once because someone suggested he believed in pure accident as the origin of living things. He said he believed no such thing. But if, like he, you rule out supernatural or intelligent causation, as a matter of method or philosophy, what are you left with?

'I wonder if he meant that the regular laws of physics and chemistry could, by themselves, lead to Life?' Mused Alisha.

'Maybe. But even if that were true, it wouldn't explain where the regular laws of physics came from in the first place. It is, in any case, very clear by now that unguided physical laws + mere chance will *never* lead to the birth of living things, even of the most simple kinds. Remember, *"it is more likely that you and your entire extended family would win the state lottery every week for a million years than for a [basic] bacterium to form by chance!"* [100]

'Science, Ali, is supposed to inquire into what is *true*. It is not science's job to shore up the materialist *belief* system and the provenly untenable belief in pure chance as a supreme creative principle. It is unfortunate, too, that so many now equate *rationality* with materialism and *materialism* with science. As though to be 'rational' and 'scientific' one must also be a materialist. Yet most of humanity's greatest thinkers and scientists have *not* been materialists. Is it true that all those great thinkers, from Pythagoras and Plato to Galileo, Newton and Planck, who were *none* of them materialists, were irrational?

'What if materialism *isn't true?* Is it then still 'rational' to believe in it? Our current science culture urgently needs to remember that: Science does not = materialism, and, materialism does not = rationality. A *true rationalist* is someone who considers the evidence and applies their logic, reason, intuition and commonsense to it at every stage.

'Life and Mind are deep mysteries, Ali. No one who can think, at least a little, can deny this. But the materialist who says, "Only physical forces are real, there *cannot* be anything more," is no more scientific than the religious person who says, "Subtle Spiritual Forces do it all, so let's not bother trying to work out *how* they do so."

'Neither of these positions is scientific. Why not? Because the first is question begging and the second is lazy and incurious.'

We walked on.

---

[100] Eastman, M & C. Missler. 1995. *The Creator Beyond Time and Space.* Emphasis added.

13 Fabulous Bats

We paused, resting in the shade of a tree by the river, before resuming our conversation. 'Biologist Michael Denton tells us that,

'The difficulty of envisaging how [hypothetical] evolutionary gaps [in the fossil records] were closed [by Darwinian processes] does not stop with birds: take the case of the bats. The first known bat which appeared in the fossil record some 60 million years ago had *as completely developed wings as modern forms.* As in the case of birds, how could the development of the bats' wings and capacity for powered flight have come about gradually?[101]

'Bats, Alisha, like other species, appeared *abruptly* on our planet.'
'No fossil ancestors? No un-bat-like species on the way to becoming bats?'
'No. By contrast with the inventive, Darwin inspired, drawings in children's and older students books, there are *no* transitional fossils of small, wingless ancestor rodents gradually changing, via a large number of intermediate species, into fast flying, upside down hanging, echo locating bats. There are the usual confident assertions that such transformations 'might' or 'must' have happened but never any convincing evidence.
'As Dr Anthony Latham says:

'What is not in any way explained [by Darwinism], however, are . . . the **sudden appearance** of completely new forms and structures [single cells, 3.8 billion years ago, then, the many amazing creatures of the Cambrian explosion, followed by amphibians, dinosaurs, birds, mammals, bats, wales, apes and people] – indeed, *all* the appearances of the phyla and classes (higher taxa). These **remain a total mystery and do not fit in with any known mechanism or theory.**'[102]

---

[101] Denton, M. 1985. *Evolution: A Theory in Crisis.* Burnett Books. p. 213. Emphasis and text in square brackets added.

[102] Latham, A. 2005. *The Naked Emperor: Darwinism Exposed.* Janus, p. 38; emphasis and words in square brackets added.

'But a lack of theory to account for the appearance of new species doesn't mean science should stop, does it?'

'Of course not. But why not revisit the classical idea that those changes may have been intelligently driven? Why is it such a taboo for our sciences to rediscover the classical understanding that living things come from subtle spiritual or super natural sources, of an astonishing intelligence and exquisite design? Doesn't this open our scientific inquiries up? Isn't it interesting?

'We would laugh at our desert islanders if they insisted that the never before seen Car, newly arrived on their island, must have a purely 'natural' or 'dumb-chance' kind of explanation.' But, when it comes to living things, this is how our current science behaves.

'So what if some things have explanations beyond their immediate physical components? Where's the problem? Isn't this a fascinating re-discovery? It's what most of the world's greatest thinkers have always believed. It's only in the last few centuries that the grim spirit of *materialism,* what the Austrian philosopher, Rudolf Steiner, called *Ahriman,* has gained such a dismal and irrational sway over the western scientific mind, leading some of our brightest intellectuals to make such patently incoherent comments as,

'We're all zombies. Nobody is conscious.'[103] or,

'You, your joys and your sorrows, your memories and free will, are in fact no more than the ['purposeless'] behavior of a vast [accidental] assembly of nerve cells and their associated molecules. As Lewis Carroll's Alice might have phrased it: "You're nothing but a pack of neurons."[104]

'It causes Marvin Minsky, an expert in *artificial* intelligence, a field whose very name derives from the existence of *natural* intelligence, to say, 'The brain is just a computer made of meat.'[105] An incredibly powerful computer which, unlike all others, arose due to *mindless* chemical coincidences! Don't people see how illogical this way of thinking is? Don't they see how it *damages* science and us alongside?

'In *Billions of Missing Links: A Rational Look at the Mysteries Evolution Can't Explain,* G.S. Simmons explains that there is *no evidence* that sonar

[103] Dennett D. 1992. *Consciousness Explained.* Back Bay Books, p. 406.

[104] Crick F. 1995. *The Astonishing Hypothesis.* Scribner, p. 3, text in square brackets added.

[105] Minsky M. Quoted in: Michalowski S. "Science, Man and the International Year of Physics." OECD Global Science Forum. Minsky of the Massachusetts Institute of Technology, cognitive scientist and artificial intelligence expert.

guided Bats evolved out of earlier, non-flying creatures. There's no evidence for *any* plausible predecessor species which gradually became bats. He writes,

> 'Biologists speculate that the bat's ancestors might have been tree dwelling shrews or moles who had some capacity to glide, but that's *mere conjecture.* It doesn't explain *echolocation,* which is their sonar ability to find food on the wing at night, or how they can purposefully drop their body temperatures at night to save energy. Most mammals constantly burn energy to maintain a specific temperature, 24/7.
>
> ... compared, by size, to our sonar detectors, theirs is *one trillion times more efficient.* Bats can fly at speeds up to 60 mph and at heights of 10,000 feet. Often they live in caves by the millions. Nurseries may have as many as 2000 pups per square meter, yet mom always seems to know her offspring.
>
> One interesting capability that breaks the rules of [Darwinian] evolution is the bat's ability to lock onto a rock and hang upside down without falling. This is made possible by an *automatic locking mechanism in their feet.* A tendon closes the toes against the rock, and then the bat's weight creates the proper tension downward to keep it locked. The moment the bat grabs the rock the clamping mechanism snaps in place.
>
> The *unprecedented* backward facing knees with forward facing feet aid the process. Bats will sleep and carry on most of their necessary activities in a hanging position for hours, only flipping 180° whenever there's a need to evacuate their bladder or defecate. One might feel sorry for the predecessors who had not yet learned to flip over.
>
> When it's time to hunt for food they just they just release, drop, and take off. Partially-hanging, sometimes-falling intermediates did not exist.'[106]

'Unfortunately, Darwin's ideas have become rather like the amusing children's story, *The Emperor's New Clothes.* Two charlatans pretended they were weaving the emperor a magnificent robe out of the richest possible golden threads, but, they maintained, those who were stupid would not be able to see it. The 'weavers' made off with all the expensive gold thread they had ordered, the Emperor was paraded around in his 'new clothes' and no one said anything for fear of being thought stupid. Until, finally, a child spoke up and pointed out he wasn't wearing anything.

'But, so far, most people in academia, lacking the intellectual courage of a Dembski, Meyer or Behe, are unwilling to speak up, as it can be risky to career or reputation to do so. Yet, gradually, more people are becoming skeptical. In

---

[106] Simmons, G. 2007. *Billions of Missing Links.* Harvest House; pps. 75–76, emphasis and text in square brackets added.

this case, the Emperor's non-clothes include Darwin's proclamation that due to macro evolution creatures would be slowly turning into other creatures,

'by differences not greater than we see between the natural and domestic varieties of the same species at the present day. [Therefore] the number of intermediate and transitional links, between all living and extinct species, must have been INCONCEIVABLY GREAT. But assuredly, IF THIS THEORY BE TRUE such have lived upon the earth.'[107]

'And, consistent with this line of reasoning, he wrote:

'. . . The number of intermediate varieties, which [if he was right] have formerly existed on the earth, (must) be TRULY ENORMOUS. Why then is not every geological formation and every stratum full of such intermediate links? Geology assuredly does not reveal any such finely graduated organic chain; and this, perhaps, is the most obvious and GRAVEST OBJECTION which can be urged against my theory.' [108]

'Darwin's admission is a major understatement. Because, far from furnishing evidence for an "inconceivably great" number of links "between all living and extinct species," the fascinating fossils provide no genuine links at all.'

'NO TRANSITIONAL SPECIES showing evolution in progress between two stages HAS EVER BEEN FOUND . . . evolution has become *an act of faith* rather than a functioning science.'[109]

'Lyn Margulis, a distinguished biologist, agreed that Darwinian thinking was inadequate to explain natural history. She noted,

'Natural selection eliminates and maybe maintains, but it doesn't create'[110]

'Expressing her skepticism, Margulis wrote that,

'. . . the Darwinian claim to explain all of evolution is a popular half-truth whose *lack of explicative power* is compensated for only by the religious ferocity of its rhetoric. Although random mutations influenced the course of evolution, their influence was mainly by *loss, alteration, and refinement.*

---

[107] Darwin, C. 1859 *On the Origin Of Species,* Ch. X, *"On the Imperfection of the Geological Record".*

[108] Ibid., Ch. IX, emphasis added.

[109] R. Milton, *Shattering the Myths of Darwinism,* (Inner Traditions Bear and Company, 2000), back cover; emphasis and text in brackets added.

[110] *Discover Magazine,* p. 68 (April, 2011).

One mutation confers resistance to malaria but also makes happy blood cells into the deficient oxygen carriers of sickle cell anemics. Another converts a gorgeous newborn into a cystic fibrosis patient or a victim of early onset diabetes. One mutation causes a flighty red-eyed fruit fly to fail to take wing. Never, however, did that one mutation make a wing, a fruit, a woody stem, or a claw appear. Mutations, in summary, tend to induce SICKNESS, DEATH or deficiencies. NO EVIDENCE in the vast literature of heredity changes shows unambiguous evidence that random mutation itself, *even with geographical isolation of populations,* leads to speciation. Then how do new species come into being?'[111]

'In an interview she said,

"This is the issue I have with neo-Darwinists: They teach that what is generating novelty is the accumulation of random mutations in DNA, in a direction set by natural selection. If you want bigger eggs, you keep selecting the hens that are laying the biggest eggs, and you get bigger and bigger eggs.

"But you also get hens with defective feathers and wobbly legs. Natural selection eliminates and maybe maintains, but it doesn't create . . . neo-Darwinists say that new species emerge when mutations occur and modify an organism. I was taught over and over again that the accumulation of random mutations led to evolutionary change – led to new species. I BELIEVED IT UNTIL I LOOKED FOR EVIDENCE."[112]

'Cautions and doubts about Darwin's ideas are not new. Even Darwin himself wrote that he was:

'well aware that there is scarcely a single point discussed in this volume on which facts cannot be adduced, often apparently leading to conclusions *directly opposite* to those at which I have arrived. A fair result could be obtained only by fully stating and balancing the facts on both sides of each question, and this cannot possibly be done here.'[113]

---

[111] Margulis, L. & D. Sagan. 2003. *Acquiring Genomes: A Theory of the Origins of the Species.* Basic Books, p. 29. Emphasis added.

[112] Quoted in "Discover Interview: Lynn Margulis Says She's Not Controversial, She's Right," *Discover Magazine,* p. 68 (April, 2011). Emphasis added.

[113] Darwin, C. 1859. reprint of 6th edition (John Murray, 1902). Emphasis added.

'Perhaps, then, it's about time more effort was put into producing this 'fair result' by more scientists doing as Darwin suggested. A small step in this direction was taken when a highly regarded Darwinian, Ernst Mayer, wrote,

'It must be admitted, however, that it is a considerable strain on one's credulity to assume that finely balanced systems such as certain sense organs (the eye of vertebrates, or the bird's feather) could be improved by random mutations.'[114]

'A rare, but refreshing, admission. So why not allow ourselves now, after so many weary years of "believing the impossible" to revisit the classical view that there may, after all, be meaning and teleology to living things? That they really *cannot* be reduced to material causes alone? They may, after all, have formal and final causes – shaping ideas and purposes, as Aristotle always said.

'In *Natural Selection or Natural Creation? – A Complete New Theory of Evolution,* John Davidson, a biological scientist who has no difficulty with the ancient yogic idea of evolution understood as meaningful change over time – of an eons long Creation in Intelligent Evolution, an ELCIE, – highlights fundamental flaws in Darwinian thinking. He writes,

'What happens, then, if one attempts to understand this vision of nature's wholeness in terms of conventional [Darwinian] evolution theory...? One is immediately led to wonder how such wholeness could have arisen from linear and random processes? ...

'Let us take a specific example: the evolution of a spider. The question is: how could a spider have evolved by random mutation? A spider exhibits a host of integrated biochemical, physiological, neurological, and anatomical, behavioral and instinctive adaptations. It is able to manufacture perfect silk; to spin and weave it into perfectly organized webs and traps; to know how to walk in a sticky web without getting entangled; to know how to wrap up a fly, but to identify and leave a wasp alone; to understand the importance of waiting patiently at the edge or center of the web; to paralyze prey with so perfect a toxin that it does not then poison itself when it sucks out the body juices with specialized mouth- parts – and so much else besides . . .

'So, musing upon how a spider could have evolved so many integrated skills and biological systems from some unspiderly ancestor, are we to suppose that the earliest spiders wove rather poorly structured webs out of inferior silk; got stuck occasionally in their own webs; could not distinguish between dangerous hornets and houseflies; used poor toxins which only

[114] Mayr, E. 1942. *Systematics and the Origin of Species.* Dover Publications, p. 296.

partially killed or paralyzed their prey while later giving them a tummy ache or worse; and that they were ill adapted for actually eating and digesting their prey and so they only relied partially upon their spiderly ways for making a living, anyway? In short, that they were pretty much of a mess, still needing considerable time, not to mention several million "chance" mutations, to get their act together?'

'Davidson continues:

'Since such evolution is still supposed to be in progress, *why are there no semi-evolved creatures living now?* Why are there *none found in the fossil record?* In fact, 300 million-year-old spiders with web-weaving spinnerets have been found in the fossil record, *yet nowhere has any fossilized species been found which is on the way to becoming a spider*. It's spiders or something else! All creatures, not just spiders, are well structured wholes. . . .

Evolution by [mere] chance and by miniature steps would pull the creature in a host of different directions until all life forms became like string and cello-tape constructions. A bit of this and a bit of that. *For one cannot attribute any goals [direction or progressive quality] to a materialistic view of evolution*. Nature cannot be assumed to know where she's going in such a[n irrational] philosophy of chance and chaos. And in the absence of any consciousness, vision or volition, how could such perfect designs as we see around us have come into existence? Yet *somehow we are asked to believe* that living creatures, who most clearly possess *volition, desires, goals, mind and consciousness,* have arisen by random processes from *dust and water* which exhibit none of these.

Materialistic evolutionary theory is forced to assume that there is no apparent purpose [teleology] to life and that natural selection is not selecting for some future, intelligently conceived goal; it is a selection only conferring advantages in the current moment. How then could some radically new species design emerge purely by chance and over an immense span of time, yet exhibit such a *multitude of integrated characteristics?* 'Progress' by tiny random steps would surely be one of constant indecision and change of direction, leading to imperfection and destruction rather than perfection and integration?' [115]

'In 1998, molecular biologist Michael Denton wrote *Nature's Destiny: How the Laws of Biology Reveal Purpose in the Universe.* He wrote,

---

[115] Davidson, J. 1992. *Natural Creation or Natural Selection – A Complete New Theory of Evolution.* Element; p. 13.

'The design of living systems, from an organismic level right down to the level of an individual protein, is so integrated that most attempts to engineer even a relatively minor functional change are bound to necessitate a host of subtle compensatory changes. *It is hard to envisage a reality less amenable to Darwinian change via a succession of independent undirected mutations altering one component of the organism at a time.*' [116]

'This is crucial. Countless, highly integrated, biochemical changes would be needed for Darwinian evolution to take place. But, there is no evidence for them. Darwin's attempts to elucidate natural history were too much influenced by his materialist philosophy. But, if science becomes overly associated with a particular belief system, like materialism, then any evidence which tends to contradict the belief system may be dismissed. For example, the well known biologist, and materialist, Richard Lewontin, writes,

'Our willingness to accept scientific claims that are *against common sense* is the key to an understanding of the real struggle between science and the supernatural. We take the side of science *in spite of the patent absurdity* of some of its constructs... in spite of the tolerance of the scientific community for *unsubstantiated just-so stories,* because we have *a prior commitment,* a commitment to MATERIALISM. ...

'Moreover, that materialism is absolute, for *we cannot allow a Divine Foot [intelligent causation] in the door.* ...To appeal to an omnipotent deity is to allow that at any moment the regularities [the regular laws] of nature may be ruptured, that Miracles may happen.' [117]

'Ok, but science can only be in a "struggle with the supernatural," as Lewontin puts it, if we know the supernatural to be unreal,' said Alisha. 'If materialism is false, then by denying all evidence for intelligence and design in nature, and for the supernatural in general, science is in a struggle with *reality,* with what it is science's very job to inquire into.

'In any case, can't the "regularities [the laws] of nature," he refers to, stem from an overarching intelligence or intelligences, like that or those to which humanity gives its traditional religious names? Cannot nature's laws themselves be seen as miraculous?'

---

[116] Denton, M. 2002. *Nature's Destiny: How the Laws of Biology Reveal Purpose in the Universe.* Free Press. p. 342

[117] Richard Lewontin (1997) Billions and billions of demons (review of *The Demon-Haunted World: Science as a Candle in the Dark* by Carl Sagan, 1997). *The New York Review,* January 9, p. 31.

'I agree, Ali. The laws of nature are not violated by the supernatural but are generated by it. And, while commonsense is not always a good guide to making sense of things, can it play *no* part in modern science? The testable, commonsense hypothesis, for example, is that no CODING system or complex life could ever arise by mere luck, no matter how much time you allow.

'Equally, could dinosaur lungs really just randomly mutate – due to rare genetic copying errors – into uniquely different birds' lungs without fatal results to the creatures concerned? "Death within minutes," as Michael Denton notes. No, that hypothesis doesn't work either, and there's no evidence for it.

'Could the crawling caterpillar really just randomly morph into the beautiful butterfly – for no reason or point at all? Again, it's implausible and, more importantly, there's no evidence for any such process. Could the land-running shrew really transform into a sonar guided bat, necessitating millions of biochemical changes, and countless transitional forms, not just one or two?

'Is there *any* real-world evidence that mere chance can create anything remotely clever? No, there is none! Is there *any* evidence that land living bears could morph into deep-sea-diving, echo-locating whales, as Darwin fantasized? No, none. Do such claims have any rational or evidential basis? No, none. Can commonsense no longer play *any* part in science?

Growing up, Ali, we were told that science is always ready to revise its ideas in light of new evidence or new thinking. Yet, despite the lack of evidence, we are still tied to Darwin's tired old steam age notions even though they are not borne out by the fossils, even though they cannot account for Life's complex DNA CODES, even though, as we'll soon see, they clash with the evidences from genetics and embryology, even though they have nothing useful to say about key ontological aspects of deep reality, like *mind and intelligence, feeling and thinking, love and desire, imagination and creativity.*

'No, we are asked to believe, post–Darwin, that Life and Mind on our planet were explained away as nothing more than mindless coincidences of mindless matter, despite this being demonstrably impossible, despite this being an offense to commonsense, despite there being no ordinary let alone any extraordinary evidence for this (oh so) very extraordinary claim.

'Yet, presently, so strange are the times in which we live, none but the most secure scientists dare, publicly, to point out that *"the Darwinian Emperor has no clothes!"* and to call out the utter poverty of the materialist philosophy which dominates our science culture today, when it comes to our thinking about origins, about the deepest questions of all, for fear of damage to career or reputation.'

We stopped talking and rested under the shade of a beautiful spreading tree. Lulled by the sounds of the nearby river we fell asleep.

14 Deceptive Icons

After our rest we resumed our walk. 'Ever since the publication of his famous work, Ali, *On The Origin of Species,* many young people, and the wider public, have been led to believe that Darwin's ideas are well and truly proven: beyond all reasonable doubt. However, in *Icons of Evolution: Science or Myth?,* Jonathan Wells carefully examines the surprisingly few items which Darwinians have had to hand to try to support their shaky theory and shows how, shockingly, they are all either misleading or false or fakes.

'So we've all been misled by Darwinism for a rather long time now?'

'Yes. Because Darwin's once seemingly interesting ideas do not stand up to close scrutiny. We've already met some of the well known icons of evolution.

'For example, Miller and Urey's famous origins-of-life experiment, which generations of young people, even practicing scientists from other fields, have been misled into believing was evidence that Life might have arisen by pure luck, when, it's now clear (a) this would be impossible (b) all the experiment produced was a few amino acids, (c) had it correctly simulated the presence of oxygen in the earth's early atmosphere, rather than excluding it, it would not have produced any building blocks at all and (d) random piles of building blocks are never known to build anything functional!

'Next, Darwin's famous Tree of Life. If mainstream biology textbooks were to be more accurate, they would explain that Darwin's notion of microbe-to-wonderful-tree-of-Life remains a hypothesis, not a well proven theory. In fact, even some of Darwin's supporters are starting to maintain that the evolution they *now* believe in is not really tree-like at all but network-like.'

'So it's not really Darwinian, then?'

'No, nor does it support Darwin's key concept of common descent. In an article, titled 'Demolishing Darwin's Tree,' the online magazine, *Evolution News (EN)* comments on a paper by researchers who are: 'ready to provide a more "expansive" view of evolution that replaces Darwin's tree with a "network" of life. Why is this necessary? Because "genetic data are not always tree-like."' Eric Bapteste's paper, 'Networks: Expanding Evolutionary Thinking,' published in *Trends in Genetics,* seeks to place evolutionary thinking within a network model. But, *Evolution News* asks, 'if they replace the tree with a complex set of inter-connections, what happens to the [iconic Darwinian] notion of universal common descent?'

"However, the more we learn about genomes *the less tree-like we find their evolutionary history to be,* both in terms of the genetic components of species and occasionally of the species themselves."

'*EN* comments: 'Interesting: if "evolutionary history" is not tree-like, does universal common ancestry still hold?' Adding, 'What matters is the extreme paradigm shift this represents. . . . Calling it "historic," the authors recognize the extent of the shift they are proposing:

'... it is a shift away from strict tree-thinking to a more expansive view of what is possible in the development of genes, genomes, and organisms through time.'

'*Evolution News* reflects:

'They use "development ... through time" as a synonym for evolution. But what kind of evolution? If it is *not* tree-like, what is it? In a network diagram, common descent gets scrambled... The new picture is of interconnected nodes, with *no clear progression from simple to complex. ... There's nothing here about a beginning and a progression...*[118] *[it's not evolutionary in the sense Darwin meant.]*

'So there's nothing really Darwinian about this proposal? It's not linear, like Darwin's theory, is it?' Commented Ali.

'No. It does away with Darwin's tree and it confirms biochemist Michael Behe's comment that, 'Darwin's idea of a "tree of life" – where a single primordial cell gave rise to all subsequent organisms – is dead. The DNA sequence data cannot be made to fit with the idea.'[119]

'Remember, the geologically *sudden* appearances of dozens of new animal forms in the Cambrian Explosion belies Darwin's idea of a gradual evolution. The Cambrian big bang radically contradicts Darwin's bottom-up idea of evolution where just one or two early anatomical designs might slowly give rise to more and more animal designs.

'Yet, Wells describes how textbook after textbook assumes common ancestry to be a proven fact and does not mention how the Cambrian Explosion makes a nonsense of this 'fact.' The Cambrian Explosion, alone, turns Darwin's entire macro evolutionary scheme upside down.

'Another famous icon relates to homology.'

---

[118] Demolishing Darwin's Tree: Eric Bapteste and the Network of Life. *Evolution News.* 9.9.13. Emphasis and text in square brackets added.

[119] Johnson, P. 2010. *Darwin on Trial.* IVP Books, p. 17.

'Homology?' Queried Ali.

'It refers to the striking similarities in skeletal structures shared by vertebrates, for example, as can be seen in the forelimbs of bats, porpoises, horses and people, including the common five finger pattern. Prior to Darwin, such similarities were seen as design archetypes. For Darwin, though, they were not evidence for design but for common descent.

'Yet, Wells explains, most text books fail to point out that the two mechanisms proposed to account for these similarities – similar genes and similar developmental pathways following conception – fail. They do not support Darwin's ideas and they do not point towards common descent. They are inconsistent with the evidence and, logically, they point to separate lines of descent, which points to intelligent design, which Darwin rejected.

'Although Darwin did not know about genes, the modern idea is that these similarities either arise due to *similar genes* inherited from a common ancestor or due to similar growth pathways following conception.

'Therefore, if it could be shown that similar structures in two different creatures are produced by similar genes this would support Darwin's idea of shared descent. 'But,' Wells writes, 'this is *not* the case and biologists have known it for decades. All the way back in 1971 Gavin de Beer wrote:

> 'Because homology [similarity of form] implies community of descent from . . . a common ancestor it might be thought that genetics would provide the key to the problem of homology [skeletal similarities]. This is where the worst shock of all is encountered . . . [because] characters controlled by identical genes are not necessarily homologous [similar] . . . [and] homologous structures need not be controlled by identical genes.'[120]

'Equally striking is that not only do *different* genes give rise to *similar* structures but, surprisingly, *similar* genes can give rise to *different* structures.

> 'Geneticists have found that many of the genes required for proper development in fruit flies are similar to genes in mice, sea urchins and even worms. In fact, gene transplant experiments have shown that developmental genes from mice (and humans) can functionally replace their counterparts in flies.' [121]

'Darwinians argue that similar biological structures arise due to a shared evolutionary history and a shared ancestry. And this idea, if true, ought to be reflected in similar growth pathways following conception. But, once again,

---

[120] Wells, J. 2000. *Icons of Evolution.* Regnery. p. 73.

[121] Ibid. p. 74.

Darwin's theory *doesn't* fit the data. Biologists have also known this for a long time. 'It is a familiar fact,' wrote embryologist Edmund Wilson, in 1894, 'that parts which closely agree in the adult, and are undoubtedly homologous [similar], often differ widely in larval or embryonic origin either in mode of formation or in position or in both.' Wells writes that,

'More than sixty years later . . . Gavin de Beer agreed: 'The fact is that correspondence between homologous structures [in the adult forms] cannot be pressed back to similarity of position of the cells in the embryo, or of the parts of the egg out of which the structures are ultimately composed, or of developmental mechanisms by which they are formed.'[122]

'Wells continues, 'The lack of correspondence between homology and developmental pathways is true not only in general, but also in the particular case of vertebrate limbs. The classic examples of this problem are salamanders. In most vertebrate limbs, [embryonic] development of the digits proceeds from posterior to anterior [tail to head] ... This accurately describes frogs but, their fellow amphibians, salamanders, do it differently . . . in the opposite direction, from head to tail.' There are other anomalies,

'Skeletal patterns in vertebrate limbs initially form as cartilage, which later turns into bone. If the development of vertebrate limbs reflected their origin in a common ancestor, one might expect to see a common ancestral cartilage pattern early in vertebrate limb development. But this is not the case. Cartilage patterns correspond to the form of the adult limb from the beginning, not only in salamanders, but also in frogs, chicks and mice. According to British zoologists Richard Hinchcliffe and P. J. Griffiths, the idea that vertebrate limbs develop from a common ancestral pattern in the embryo 'has arisen because investigators have **superimposed their preconceptions'** on the evidence. [123]

'Yet, as Wells explains, many biology textbooks fail to point out that the two mechanisms proposed to account for the similarities in vertebrate limbs do not support Darwin. Wells concludes this part of his book by asking if the continuity of information required to build up vertebrate limbs neither comes from genes nor developmental pathways, how can it come from common descent?

'While the evidence points to separate lines of descent, this conclusion is disregarded. Why? Because it conflicts with Darwin's theory.

---

[122] Ibid. p. 71.

[123] Ibid. p. 72–73, emphasis added.

'Another famous Darwinian icon is Ernst Haeckel's idea that 'ontogony recapitulates phylogeny.'

'Meaning?' Asked Alisha.

'This is the idea that the movement of water breathing creatures, out of the sea, via an amphibious *(water-to-air-breathing)* stage, onto the land, can still be seen in the early developmental stages of contemporary embryos – including what might, at first glance, appear to be vestigial 'gill slits.'

'In other words demonstrating a fish ancestry?'

'Yes. To back up this idea, Haeckel produced a set of drawings of different embryos, including fish, salamander, tortoise, chick, rabbit and human, seeking to show how similar they all are, especially early on. Unfortunately, and to his discredit as a scientist, Haeckel doctored his drawings to make the embryos look more alike than they really are. He did this to bolster Darwin's idea of common descent. In fact, what Darwin believed to be the supporting evidence from embryology, as misrepresented by Haeckel, gave him a greater confidence in his own theory. He didn't know Haeckel's drawings were doctored and he found them very persuasive. He wrote,

> 'it is probable, from what we know of the embryos of mammals, birds, fishes and reptiles, that these animals are the modified descendants of some ancient progenitor they had in common.'

'Wells writes that since Haeckel's embryo drawings seem to 'provide such powerful evidence for Darwin's theory . . . some version of them can be found in almost every modern textbook dealing with evolution.' Yet it turns out that *biologists have known for over a century that Haeckel faked his drawings.* Vertebrate embryos never look as similar as he claimed. Wells writes,

> 'Although you might never know it Darwin's "strongest single class of facts" is a classic example of how evidence can be twisted to fit a theory.' [124]

'Wells notes that if one started with the evidence, then examined Darwin's ideas, one would likely conclude that the various classes of vertebrates, rather than being descended from a common ancestor, probably had separate origins. But Darwin's followers have always found it hard to accept the embryological evidence, so they try to mould it to match Darwin's ideas.

'Wells explains that Von Baer, a famous 19th century embryologist, couldn't support Darwin precisely *because* of the conflict with the evidence from embryology. Yet so many people rushed to embrace Darwin's ideas before taking the real-world evidence into account. Wells writes,

---

[124] Ibid. p. 83.

'Many modern Darwinists haven't changed. It doesn't matter how much the embryological evidence conflicts with evolutionary theory – the theory, it seems, must not be questioned. This is why, despite repeated disconfirmation, Haeckel's biogenetic law and faked drawings haven't gone away.'[125]

'World famous fossil expert, S. J. Gould referred to Haeckel's faked embryo drawings as *'the academic equivalent of murder,'* yet they are still around.

'Wells notes that more accurate textbooks would not recycle misleading drawings of embryos, would not call pharyngeal pouches 'gill slits,' as though these pouches are evidence of fish ancestry *(they have nothing to do with gills!)*, and would explain to their students that embryology tends to *refute,* not confirm, Darwin's ideas. Yet many textbooks still fail to explain that earlier embryonic stages are *dissimilar* and they perpetuate the misleading claim that the alleged similarities are evidence for common ancestry.

'The next icon Wells discusses is *Archaeopteryx.* Far from questioning the, by now, dubious claim that this ancient bird provides evidence for a transitional link between reptiles and birds, many textbooks give no clue that there is any controversy over its ancestry nor that modern birds are probably *not* descended from it.

'Next, the Galapagos finches. Rather than explaining that selection on finch beaks oscillates between wet years and dry years without resulting in any net evolutionary gain, many textbooks mention the *within-species* finch variations as though they are good evidence for Darwin's theory of *macro-evolution.*

'Wells goes into other Darwinian icons such as peppered moths, four winged fruit flies, evidence from fossil horses and theories about humans origins which latter are 'controversial, and they rest on little evidence; all drawings of "ancestors" are hypothetical.' As Casey Luskin points out, while Darwinians often claim that the evidence in favor of an evolution of humans from ape-like ancestors is overwhelming, in fact,

'hominin fossils generally fall into one of two categories — ape-like species or human-like species (of the genus *Homo*) – and that there is a large U N B R I D G E D G A P between them. ... *the fragmented hominin fossil record does not document the evolution of humans from ape-like precursors.* In fact, scientists are quite sharply divided over who or what our human ancestors even were. Newly discovered fossils are often initially presented to the public with great enthusiasm and fanfare, but once cooler

---

[125] Ibid. p. 101.

heads prevail, their status as human evolutionary ancestors is invariably called into question.

'The details of the earliest stages of human origins are murky. They come from what UC–Berkeley paleoanthropologist Tim White once called "a **black hole** in the fossil record."[126] ...

'While early hominin fossils are controversial, due to their fragmented condition, there is one major group – the australopithecines – that is widely promoted as directly ancestral to humans. The primary claim is that australopithecines had the head of a chimpanzee but a body that allowed it to walk upright, like humans.

Despite the prevalence of that standard view, authorities have found that the fingers, arms, chest, hand bones, striding gait, shoulders, abdomen, inner-ear canals, developmental patterns, toes, and teeth of australopithecines *point away* from their being human ancestors and/or suggest that they *didn't* have human-like bipedal locomotion.[127] [This includes the famous australopithecine referred to as 'Lucy.'] ...

Paleoanthropologist Leslie Aiello, who served as head of the anthropology department at University College London, stated that when it comes to locomotion, "Australopithecines are like apes, and the *Homo* group are like humans. Something major occurred when *Homo* evolved, and it wasn't just in the brain."'[128]

'On the Big Bang-like appearance of humanity,

'When the human-like members of our genus *Homo* appear, they do so abruptly. A paper in the Journal of Molecular Biology and Evolution called the appearance of *Homo sapiens* "a genetic revolution" in which "no australopithecine species is obviously transitional."[129] In a 2004 book, the famed evolutionary biologist Ernst Mayr explained that "the earliest fossils of *Homo*, *Homo rudolfensis* and *Homo erectus* are separated from *Australopithecus* by a large, unbridged G A P" without "any fossils that can serve as missing links."[130]

---

[126] Tim White, quoted in Ann Gibbons, "In Search of the First Hominids," *Science* (Feb. 15, 2002), 295:1214–1219.

[127] For a more detailed discussion of the fossil evidence and human origins, see Casey Luskin, "Human Origins and the Fossil Record" in *Science and Human Origins* (Discovery Institute Press, 2012), pp. 45–83.

[128] Leslie Aiello, quoted in Richard Leakey and Roger Lewin, *Origins Reconsidered: In Search of What Makes Us Human* (Anchor Books, 1993), p. 196.

[129] John Hawks et al., "Population Bottlenecks and Pleistocene Human Evolution," *Journal of Molecular Biology and Evolution* (2000), 17(1):2–22.

[130] Ernst Mayr, *What Makes Biology Unique?* (Cambridge Univ. Press, 2004), p. 198. Emphasis added.

... the fossil record provides us with ape-like australopithecines ("before") and human-like *Homo* ("after"), but not with fossils documenting a transition between them. In the absence of intermediaries, we're left with "inferences" of a transition based strictly upon the *[prior] assumption* of Darwinian evolution. No wonder one commentator argued that if we take the fossil evidence at face value, it implies a *"big bang theory"* of the appearance of our genus *Homo*. [131]

... the evidence shows that human-like forms appear *abruptly* in the fossil record, *without any fossils* connecting us to our alleged ape-like evolutionary ancestors. This contradicts the expectations of neo-Darwinian evolution and suggests that unguided evolutionary mechanisms do not account for the origin of our species.'[132]

'The comment, "if we take the fossil evidence at face value," "it implies a big bang theory of the appearance of humanity," is telling. Why would we *not* take the evidence at face value unless we have already decided that some *prior* theory, like Darwin's, is the only one possible?

'There is, in any case, much more that separates people from apes than differences in gait, shoulders, chest, inner-ear canals, teeth and developmental patterns. Take our capacities for abstract thought, language and speech.

'In his recent book, *The Kingdom of Speech,* Tom Wolfe notes that not only is the human capacity to speak unique but that, despite a vast amount of work on this, over decades, "by some of the greatest minds in academia," all evolutionary theories about the origins of human language are incredible and they fail to explain this unique feature of what it is to be human. Reviewing Tom Wolfe's book, Michael Egnor writes,

'... Darwinists are at an utter loss to explain how language... "evolved." ... [No] evolutionary just-so stories come anywhere close to explaining how man might have acquired the *astonishing [and biologically unnecessary]* ability to craft unlimited propositions and concepts and subtleties within subtleties using a system of grammar and abstract designators (i.e. words) that are utterly lacking anywhere else in the animal kingdom.

'Darwin and his progeny have had no dearth of fanciful guesses – birdsongs (Darwin's favorite theory) and grunts and grimaces that mutate... into Cicero and Shakespeare. Evolutionary theorizing about language [Wolfe notes] has been a colossal waste of time ['by some of the greatest minds in

[131] "New study suggests big bang theory of human evolution," (Jan. 10, 2000) at http://www.umich.edu/~newsinfo/Releases/2000/Jan00/r011000b.html. Emphasis added.

[132] Has Science Shown That We Evolved from Ape-like Creatures? Casey Luskin *Salvo magazine* September 26, 2013. Emphasis added.

academia' quoting Chomsky]. *None of this evolutionary fancifulness makes any sense, nor has any real scientific basis...*

'... the human mind is qualitatively different from the animal mind. The human mind has immaterial abilities – the intellect's ability to grasp abstract universal concepts divorced from any particular thing – and that this ability makes us more different from apes than apes are from viruses. We are *ontologically* different. We are *a different kind of being* from animals. We are not just animals who talk. Although we share much in our bodies with animals, our language – a simulacrum of our abstract minds – has no root in the animal world.

'Language is the tool by which we think abstractly. It is sui generis. It is a gift, a window into the human soul, something we are made with, and it did not evolve. Language is a rock against which evolutionary theory wrecks, one of the many rocks – *the uncooperative fossil record, the jumbled molecular evolutionary tree, irreducible complexity, intricate intracellular design, the genetic code, the collapsing myth of junk DNA, the immaterial human mind* – that comprise the shoal that is sinking Darwin's Victorian fable.' [133]

'Towards the end of *Icons of Evolution,* in a chapter headed *Science or Myth?,* Wells quotes famous Darwinian, Ernst Mayr, who proposed, in *Scientific American,* that no educated person doubts any longer the 'so-called theory of evolution, which we now know to be a simple fact.' Wells writes,

'Ask any educated person how we know that evolution is a simple fact... and chances are that person will list some or all of the icons described in this book. For most people – including most biologists – the icons *are* the evidence for Darwinian evolution.

'As we have seen, however, the icons of evolution *misrepresent* the evidence. One icon (Miller–Urey) gives the false impression that scientists have demonstrated an important first step in the origin of life. One (the four–winged fruit fly) is portrayed as though it were raw materials for evolution, but is actually a hopeless cripple – an evolutionary dead end. Three icons (vertebrate limbs, *Archaeopteryx,* and Darwin's finches) show actual evidence but are typically used to conceal fundamental problems in its interpretation. Three (the tree of life, fossil horses and human origins) are incarnations of [materialist] concepts masquerading as neutral descriptions of nature. And two icons (Haeckel's embryos, and peppered moths on tree trunks) are fakes.

---

[133] Language Is a Rock Against Which Evolutionary Theory Wrecks Itself. *Evolution News,* Michael Egnor. 09.19.16. Emphasis added.

'People such as Ernst Mayer insist that there is overwhelming evidence for Darwin's theory. ... [Yet,] If there is such overwhelming evidence for Darwinian evolution, why do our biology textbooks, science magazines and television nature documentaries keep recycling the same tired old myths?

'There is a pattern here, and it demands an explanation. *Instead of continually testing their theory against the evidence,* as scientists are supposed to do, some Darwinists consistently ignore, explain away, or misrepresent the biological facts in order to promote their theory. One isolated example of such behavior might be due simply to overzealousness. Maybe even two. But ten? Year after year?[134]

'In his review of *Icons of Evolution,* Dean Kenyon, biology professor, writes that Wells 'has brilliantly exposed the exaggerated claims and deceptions that have persisted in standard textbook discussions of biological origins for many decades, in spite of contrary evidence! '

'So,' mused Ali, 'the iconic 'proofs' of evolution we've been brought up on, for generations now, provide little real support for Darwin's grand theory?'

'Correct. They are either evidence for micro-evolution or they are misleading or they are fakes. In *Shattering the Myths of Darwinism,* science writer Richard Milton confirms these suspicions. He writes,

'Darwinism... totters atop a shambles of outdated and circumstantial evidence that in any less controversial field would have been questioned long ago.

Darwin's theory of evolution relies on the idea of random genetic mutation, coupled with natural selection, altering species over the course of 3.8 billion years. But all experimental evidence indicates that the extent of genetic change wrought by natural selection is quite limited. . . . [it is also an] embarrassing fact that . . . NO TRANSITIONAL SPECIES SHOWING EVOLUTION IN PROGRESS BETWEEN TWO STAGES HAS EVER BEEN FOUND . . .

[The] theory of evolution has become *an act of faith* rather than a functioning science. ... Not until the scientific method is [properly] applied to Darwinism will it be exposed, and only then will the right questions be asked about the mystery of life on earth.'[135]

We walked on.

---

[134] Wells, J. 2000. *Icons of Evolution.* Regnery. pp 229 – 230; emphasis and text in square brackets added.

[135] Milton, R. 2000. *Shattering the Myths of Darwinism.* Inner Traditions; back cover, text in brackets added.

15 Is Sudden Appearance Evolution?

'The ABSENCE OF FOSSIL EVIDENCE for intermediary stages between major transitions in organic design, indeed our inability, even in our imagination to construct functional intermediates in many cases, has been a persistent ... problem for gradualistic accounts of evolution.' S. J. Gould[136]

'DARWIN'S PREDICTION of rampant, albeit gradual, change affecting all lineages throughout time is REFUTED. The record ... speaks for *tremendous anatomical conservatism*. Change in the manner Darwin expected is just not found in the fossil record.' N. Eldredge.[137]

'In spite of these examples, it remains true, as every paleontologist knows, that most NEW SPECIES, genera, and families and that nearly all new categories above the level of families APPEAR in the record SUDDENLY and are not [as Darwin thought] led up to by known, gradual, completely continuous transitional sequences.' G. Simpson.[138]

'These statements, by world class fossil experts, all falsify Darwin.'

'Should we not take them seriously then?' Asked Alisha.

'Yes. . . but don't be misled, they all remained Darwinian in spirit.'

'Why? The fossils, held to be so key to his theory, do not support Darwin.'

'Yes, but, to their way of thinking, their statements didn't change very much. They still believed in Darwin's notion of common descent, and, governed by the constraints of methodological materialism and so called naturalism (MM/MN), they would be obliged to continue to look for a wholly materialistic theory to explain the data – even if Darwin's ideas proved false.'

'This is why Eldredge and Gould came up with a new idea which they called *punctuated equilibrium or PE*. PE tries to account for the fact that life on earth is very stable for long periods but then experiences rapid change, characterized by the geologically sudden appearances of new species. But the problem with PE, as with Darwin's idea of a much slower evolution, is that the

---

[136] Gould, S. J. 'Is a new and general theory of evolution emerging?' *Paleobiology,* vol. 6(1), January 1980, p. 127. Emphasis added.

[137] Eldredge, N. & I. Tattersall. 1982. *The Myths of Human Evolution.* Columbia University Press, p. 48.

[138] Simpson, G. G. 1955. *The Major Features of Evolution.* Columbia University Press, p.360.

often harmful effects caused by random genetic mutations, could *not*, realistically, lead to the vast changes in form and function needed to make *either* theory work.

'While the far more rapid changes called for by PE match the fossil evidence better than Darwin's ideas, those changes cannot be explained by random mutations. For Darwin, evolution needed "infinitely numerous transitional links" forming "the finest graduated steps." He believed that "nature does not make leaps."[139] Yet the vast numbers of transitional species he called for are, it must be emphasized over and over again, **never to be found** and, as Eldredge and Gould conceded, the fossils *do* make leaps.

'New species *do* materialize into physical reality in a mysteriously abrupt way, not in the very gradual manner Darwin claimed. Then, they remain the same *without* morphing into anything else. The world's fossil records illustrate the tremendous *stability* of the species, the very opposite of Darwin's ideas.

'Ironically, the fossils refute Darwin. At the same time, skillful human selectors have never been able to create a genuinely new species from a prior species. Darwin ignored this. He wanted to believe that species could vary indefinitely.

'To those who argue that it is earlier, now extinct, forms which gave rise to gradually diverging groups of creatures, there is *no evidence* for such ancestors. As Colin Patterson explained to Tom Bethel, the spaces created for them on the branching off points on Darwin's hypothetical tree of evolution remain empty. There are inventive drawings in children's books, and the claim that the shared ancestors 'must have' been there, because, without them, Darwin's theory doesn't work. But this is putting Darwin's beliefs before the evidence. It is back to front. It is doing science backwards.

'So it's unscientific to ignore the evidence?'

'Yes. We cannot just ignore the intriguing issue of *sudden appearance*.

'Isn't that odd? How could the species just suddenly appear?' Asked Alisha.

'Well, Ali, think of the Big Bang. An entire universe appearing from no-where to now-here, more or less instantaneously. Isn't that odd? It, too, is a sudden materialization. *It's not some slow, evolutionary thing.*

'Are you saying, then, that some things just . *m a t e r i a l i z e* out of the cosmic void, seemingly out of nothing?'

'Well, while we currently do not know how this works, or why, the evidence points more this way than towards the kinds of very slow, and wholly linear, processes Darwin believed in. His idea – that all creatures *now* existing

---

[139] Battson, A. *Conflicts Between Darwin and Paleontology.* http://www.veritas- ucsb.org/library/battson/stasis/2.html

might have descended from just one or two common-ancestor microbes – was worth considering. But it is not supported by the data. The absence of the vast numbers of transitional fossils needed to make his theory work, the *strict limits* to genetic change, which the world's most skillful breeders have never been able to cross, the mysteriously *jumping, not gradual,* appearances of new species, the evidences from embryology, bio-chemistry and genetics, all conflict with Darwin's ideas.

'What do you propose, then?' Queried Alisha, looking thoughtful.

'That we work with the data we *have,* not the data we'd *like* to have, be we Darwinians, who try to *force* the fossils to fit their ideas, or young-earth creationists who try to impose Bishop Ussher's 17th century ideas onto the evidence. Both groups try to make the data fit their pet theories. But the evidence is that, in conflict with bible-based, young-earth, creationism, Life on Earth is far older than the few thousand years Bishop Ussher allowed for.

'The evidence is that, in conflict with Darwin, the species *did* suddenly emerge, from seemingly no-where to now-here, in a mysterious pattern of repeating Life Waves, followed, at times, by mass extinctions. While this doesn't amount to a new theory of Life, it does free us to consider the evidence as it *is* and not continue trying to force it to fit either Darwin or Ussher's ideas.

'It's also useful to remember that in deep reality, as Max Planck the discoverer of quantum physics pointed out, *there is no matter as such.* Just
e - - - n - - - e - - - r - - - g - - - y   s - - - - p - - - - a - - - - c - - - - e
99.9 . . . 9 . . . 9 . . . 9 . . . 9 . . . 9 . . . 9 . . . 9 . . . 9 . . . 9 . . . 9 . . . 9%
of it – and everything, *this* amazing second and *every* amazing second, is disappearing and reappearing, de-materialzing and re-materializing, out of the quantum ground, now here, now not here, trillions of times a second.'

'I'm not sure what you're saying here . . .' said Alisha.

'That *all* things, Ali, are *materializations,* some of which occur extremely rapidly, as in the Big Bang, some relatively rapidly, as in the Cambrian explosion, and, others, take slower, more sequential forms.

'The key question for us, though, is not materialization, the evidences for which are all around us, from atom to galaxy and everything in between, but whether the Plants, Animals and People are, in their essence, *intelligently* caused phenomena, with random causation playing a subsidiary role, or, are they just 'Flower shaped,' 'Eagle shaped' and 'People shaped' coincidences of chemistry, as our science currently maintains, taking its anti-spiritualist lead from Darwin. Which of these two options makes more sense?'

'Ok, but couldn't Darwin be *mostly* correct, just mistaken in his belief that life could first **arise** and then **evolve** onwards from there by pure chance?'

'No, because even if, unlike Darwin, we allow for intelligent causation, there is no evidence for macro evolution as he conceived it, for one species changing, very gradually, via many transitional species, into something else.

'Take bats. Let's say a ground-based rodent was very gradually, *but intelligently,* transformed into a bat, via many transitional species, as Darwin predicted. The trouble is, there is no evidence for such transformations, random *or* intelligently guided.

'No, the new species, when they arose in the deep past, emerged in a mysterious pattern of sudden appearances, in a series of biological 'big and little bangs' as in the Cambrian. They did not give rise to new forms, but they either continued, with relatively minor changes over time, or, eventually, they disappeared into extinction.

'As to why nature takes these fascinating forms, no one can say. In Hindu thought, God or Supreme Intelligent Beingness, which *just mysteriously IS,* explores and gives expression to reality in many different ways and all living things have their own reasons for being, all experiencing and playing in the physical worlds in their own way. Perhaps this is just how it is.

'But just because we can't see the processes by which `spiritual ideas` are materialized it doesn't mean they're not possible. If the blazing, amazing, materialization of an entire universe, from no-where to now-here, was possible, then other kinds of abrupt, non-evolutionary materializations may also be possible. The issue, though, is not, fundamentally, materialization but whether life's amazing manifestations, including we, are expressions of something *intelligent and multi-dimensional* or meaningless and accidental.

'The fossil case for Darwin was poor in his own day. Today, it is even clearer. The fossils continue to disagree with Darwin. Firstly, because of,

> '***Stasis*** – most species exhibit *no directional change* during their tenure on earth. They appear in the fossil record looking much the same as when they disappear; morphological *change is usually limited and directionless.*' And, secondly, because of, '***Sudden Appearance*** – in any local area, a species does not arise gradually by the steady transformation of its ancestors; [rather] it *appears all at once and 'fully formed.'* [140]

'In fact, all the way back in 1980, S. J. Gould wrote that the modern, neo-Darwinian notion of an evolution driven by random genetic mutations was:

> 'effectively dead, *despite its persistence as textbook orthodoxy.*'[141]

---

[140] Gould, S. J. 198 , emphasis added.

[141] Gould, S. J. 1980. *Is a New and General Theory of evolution emerging?* p.120, emphasis added.

'Gould's well known colleague Niles Eldredge wrote, 'Darwin's prediction of rampant, albeit gradual, change affecting all lineages throughout time is refuted. The record ... speaks for tremendous anatomical conservatism. Change in the manner Darwin expected is just not found in the fossil record.' [142]

'Antony Latham writes, "the general situation found in successive strata is that of *stasis.*" He comments on a study of the trilobite Carolinites, by McCormick and Fortey, as it progressed for many millions of years through the Lower to Middle Ordovician period:

> 'Changes were seen over time in various characteristics of the trilobite. Some changes of anatomy were sudden, some seemed to go through successive transitional forms and some fluctuated with little or no change. Those changes that occurred showed evidence of reversal, that is the changes sometimes reverted back to the original. Statistical analysis of the entire process showed that there was *no evidence of any sustained "direction" in any changes* ...

'He adds:

> *'What is not in any way explained,* however, are the big changes that we call macro evolution these *are the sudden appearance of completely new forms and structures* – indeed, all the appearances of the phyla [basic animal designs] and classes (higher taxa). These *remain a total mystery* and do not fit in with any known mechanism or theory.[143]

'Of the echinoderms, the back-boneless (invertebrate) design to which star fish and sea urchins belong, Latham writes,

> *"there are no transitional forms* found in the fossil record that link the various types of echinoderm since their first appearance in the Cambrian period. What we see is the initial phylum [anatomical design] appearing suddenly around 530 million years ago – in a wide variety of forms . . . *I have searched hard for any evidence of transitional forms* between the major groups which exist today and I have seen NONE. What we do see is small changes in existing groups (micro-evolution). The initial phylum appears along with all the others in the Cambrian explosion, it is fully complex and has more classes than we have now; the classes that survive appear suddenly also. The only clear evolution is on the micro-scale. *It is the sudden*

---

[142] Eldredge, N. & I. Tattersall. 1982. *The Myths of Human Evolution.* Columbia University Press, p. 48, emphasis added.

[143] Latham, A. *The Naked Emperor: Darwinism Exposed. London:* Janus, p. 59 emphasis and words in square brackets added.

appearance of new forms without linking transitionals that remains the real mystery – just as it was in Darwin's day.

All that is written about the echinoderms is mirrored in the fossil record of the other invertebrates. I have used the echinoderms as one example of what is found also amongst fossils of the other main phyla of invertebrates: arthropods, mollusks, sponges, cnidarians, bryozoans and brachiopods. Check any textbook on the subject and you will see that *sudden appearance and discontinuity is the norm*. The problem that Darwin so clearly saw has not gone away with time and the *fossil record seems now to refute the entire basis of his theory of gradual changes.* "[144]

'A long time ago now, in 1979, Luther Sunderland wrote to the then senior fossil expert at the Natural History Museum in London, Colin Patterson, to ask him why, in his recently published book, *Evolution,*[145] he had not included a single photograph of a transitional fossil demonstrating evolution. Patterson replied with this interesting and refreshingly honest admission.

'I fully agree with your comments on the lack of direct illustration of evolutionary transitions in my book. *If I knew of any, fossil or living, I would certainly have included them.* You suggest that an artist should be used to visualize such transformations, but where would he get the information from? I could not, honestly, provide it, and if I were to leave it to artistic license, would that not mislead the reader?'

Patterson added:

'Yet Gould and the American Museum people are hard to contradict when they say THERE ARE NO TRANSITIONAL FOSSILS. ... You say that I should at least "show a photo of the fossil from which each type of organism was derived." I will lay it on the line—there is NOT ONE SUCH FOSSIL for which one could make a watertight argument.'[146]

'So Colin Patterson, another world-class fossil expert, like Gould and Eldredge, was willing to admit that there is no convincing evidence for Darwinian macro evolution?'

'Correct. Unlike some of his more ideological colleagues, he was true to the evidence. He was not trying to wave it away. This is, also, what Eldredge and Gould were attempting to do with their idea of punctuated equilibrium or PE.

---

[144] Ibid. p. 61 emphasis and words in square brackets added.

[145] Patterson, C. 1978. *Evolution.* The British Museum of Master Books, Natural History.

[146] Sunderland, L. 1988. *Darwin's Enigma.* p. 89. Emphasis and text in square brackets added.

'What was their idea precisely?'

'That the species will be stable for long periods without changing in the way Darwin predicted, an idea which, unlike Darwin's, *does* match the fossil evidence, but, they said, when macro-evolutionary change does occur, it is *so* rare, *so* geographically limited and of such *short* duration that *no fossil evidence* can ever be found for it.

'This, though, is odd. Rather than take the real world fossil evidence at face value, it's another attempt to account for the clear lack of evidence in support of Darwin's overall theory. "The data is 'wrong' not Darwin!"

'Remember, Eldredge and Gould were not anti-Darwinian. They were just trying to have it both ways. They admitted the fossils didn't support Darwin but, they said, he must still be right about common descent, if not about his prediction of an "inconceivably great" number of transitional species. Why?

'Because, they said, the species, when they *did* transform, from one kind into another kind, (worm-to-fish, fish-to-frog, dinosaur-to-bird, bear-to-whale, shrew-to-bat etc), did so *so rapidly, and so rarely,* that the large numbers of transitional forms predicted by Darwin's theory were not needed. Yet Eldredge and Gould had no plausible mechanism to drive the very rapid 'hopeful monster' changes they argued for. All they had was the modern 'scientific' belief that *Life on Earth 'must have' arisen and evolved by pure chance.*

'So what's the answer, Ollie? You seem to believe very firmly in sudden appearance rather than the very slow kind of evolution Darwin argued for.'

'Hold on, Ali, I don't 'believe' in sudden appearance. I believe the *facts!* Our task is to make sense of the facts: the *sudden* appearances of new species, followed by long periods of stability, which Darwin's theory does not explain.

'As to 'belief,' there are only two possible beliefs we can have about Life and Mind on Earth. Either they have random causes or intelligent. Materialism rules out intelligent causation because Life's deeper causes cannot be seen physically. Well, that's blindingly obvious. If it wasn't so there'd be no debate.

'However, we can infer a cause from its effects. Landscapes can, reasonably, be explained in terms of random, unintelligent design. But living things, vastly more complex than anything we make, cannot. It's as simple as that. This, though, is where *faith* in the materialist belief system comes in. It's easy to show that Life could never either arise or evolve by mere luck. But the Darwinian true *'believers,'* the *'faithful,'* take no notice.

'Darwin's theory of Life fails twice. Firstly, because it is not supported by the fossil and other evidence. Secondly, even if it was supported by the fossil evidence, it would still fail. Why? Because it is easy to show that the kind of macro-evolution he argued for, even if real, could only take place if it was *intelligently* guided. It was far too complex to take place by 'accident.'

16 No One to Know, Nothing to Know

With the river at our side, we continued our conversation. 'The Life forms of our planet, Ali, appear in a deeply mysterious way, not at all as predicted by Darwin. In *Evolution News,* Casey Luskin writes, [147]

'The fossil record's overall pattern is one of *abrupt explosions* of new biological forms, and generally lacks plausible candidates for transitional fossils, contradicting the pattern of gradual evolution predicted by Darwinian theory.

'This non-Darwinian pattern has been recognized by many paleontologists [fossil experts]. University of Pittsburgh anthropologist Jeffrey Schwartz states: "We are still in the dark about the origin of most major groups of organisms. They appear in the fossil record as Athena did from the head of Zeus — full-blown and raring to go, in contradiction to Darwin's depiction of evolution as resulting from the gradual accumulation of countless infinitesimally minute variations."[148]

'Likewise the great evolutionary biologist Ernst Mayr explained that "new species usually appear in the fossil record suddenly, not connected with their [alleged] ancestors by a series of intermediates." [149]

'Similarly, a zoology textbook observes: "Many species remain virtually unchanged for millions of years, then suddenly disappear to be replaced by a quite different, but related, form. Moreover, most major groups of animals appear ABRUPTLY in the fossil record, FULLY FORMED, and with no fossils yet discovered that form a transition from their [alleged] parent group."[150]

'What a mystery is existence! But is it intelligent or meaningless? The modern idea that it's a 'dumb' mystery, that everything in existence can be assembled by pure chance, by 'unintelligent design,' is baffling to me, Alisha.

---

[147] Casey Luskin, March 10, 2015, *Evolution News,* Information for Students about the Scientific Dissent from Darwinism List. Emphasis added.

[148] Jeffrey Schwartz, *Sudden Origins: Fossils, Genes, and the Emergence of Species,* p. 3, Wiley, 1999.

[149] Ernst Mayr, *What Evolution Is,* p. 189, Basic Books, 2001.

[150] C.P. Hickman, L.S. Roberts, and F.M. Hickman, *Integrated Principles of Zoology,* p. 866, 1988, 8th ed. Emphasis and text in square brackets added.

'What do we imagine is causing our minds to think, our hearts to beat and our lungs to breathe, right now? Does anyone *really* believe it could all be due to 'purposeless' chemical luck? Does anyone *really* think that our abilities to imagine and to feel, to create art and to discover mathematics are the *unintended byproducts – mere epiphenomena – of insensate atoms* of carbon, hydrogen and oxygen being tossed around in a mindless cosmic ocean until they randomly bump together to become microbes and, eventually, *living, feeling, thinking* people, like you and me, all for no reason or point at all?

'Darwinian thinking has no real explanation as to why Life would ever arise or evolve by pure chance. There are *no rational – and zero evidential –* grounds for claiming that mere luck could drive an unfoldment of life in a progressive way of ever rising "improvement" and increasing intelligence, leading, eventually, to the human mind and its capabilities, other than: 'Now we realize the world wasn't, literally, made by a Father Christmas type god, out of spit and clay, in just a few busy days, what other option is there?'

'Contemporary ID thinkers, like Behe, Meyer and Dembski, exploring modern theories of intelligent design versus Darwin's theory of *unintelligent* design are not giving up on science, Ali. They're not religious literalists who try to treat the bible, or any other religious texts, as 'science.' They're just as keen as anyone else to explain reality and for science to go as far as it can.

'All they're saying is that just because it's relatively easy to explore physical nature and more demanding to inquire into nature's more subtle causes, we shouldn't give up and assume that Life and Mind *have* no deeper causes. It's the current mainstream rejection of all ideas of the *super* natural and the *spiritual,* which drives the world's famous materialists, Dawkins et al.

'But they are way too dogmatic with their materialism. I'd love say to them,

> 'Look, you don't have to belong to this or that religion, or believe everything was made by a risibly simplistic, Father Christmas type god, out of spit and clay, in just six literal days. But, can you not conceive that all of it, from quark to galaxy, might, after all, be an expression of something *ontologically intelligent,* whether we refer to it as God or the Tao or simply the 'Force,' the name doesn't matter, rather than mindless *(without Soul),* dead *(without Life Forces)* and completely irrational?
>
> 'Is it really logical to argue that our *sensitive feelings* come out of *dusty, dead matter, by pure accident?* Is it really rational to claim that our *bright, intelligent minds, 'unintelligently' and 'purposelessly' emerge from mindless cosmic dust, in pointless and accidental combination,* descended by 'chance,' *(by what chance!?),* from some supposedly accidental microbial cell, itself more complex than a modern spaceship and supercomputer

combined, than that they meaningfully emerge from a *multi-dimensional causal matrix* which, to put your minds at rest, looks nothing like Father Christmas! Wouldn't that be a welcome release from your depressing paradigm of utter unmeaning?'

'But where would such an *ultimate, intelligent, Energy, Law, Force or Source* come from? Who'd make it? Who'd design it?' They say, echoing a comment Richard Dawkins once made, "Who 'designed' God, then?"

'No one! It, whatever labels we may like to give it, just Is. Who, after all, designs all the random, pointless happenings you fill your knowledge–gaps with and which you believe to be so extraordinarily creative?'

'No one, *they just pointlessly and mindlessly Are.'*

'But, where do they come from? What *causes* them? They must have a cause of some kind.'

'Well, they must come from some kind of Mindless, Unintelligent and Accidental Universe Generating Mechanism (MUA-UGM) which *just IS.'*

'Ok... but where does your MUA-UGM come from? *Who or What caused it to exist, to so mysteriously, if pointlessly, just be?'*

'Who knows?! It just IS!' They say. 'Beyond the MUA-UGM, required to generate all the mindless random events, which eventually lead – by pure, stupid chance – to nature's incredibly precise laws, her smart materials, to the fascinating Living World, to our own curious, inquiring Minds, and to our sometimes caring Hearts, there is no useful or possible regress.'

Alisha smiled and said, 'So this MUA-UGM functions as a logically necessary pseudo-god, a final source and uncaused cause, which just mysteriously is, for the modern atheist and materialist belief systems?'

'Yes. Because all manifested things have causes. While we may never know why there is anything at all rather than nothing at all, we are, surely, capable of working out whether our ultimate cause or causes is or are more likely to be mindless or intelligent? Today's science maintains that we accidentally emerge from the *de-intelligenced and de-Souled matrix of its dry faith,* imbued with consciousness and full of purposes and desires but that, really, our curious, creative minds, our desires and our loves, our ideas and our plans, are *all just accidents of mindless cosmic dust.* This is held to be the 'respectable,' the 'scientific' view. That it is illogical and, quite literally, *impossible* is ignored.

'It does seem unrealistic.' Said Alisha.'

'That's the point, Ali, it's unreasonable. The trouble with materialism's MUA-UGM is that it's intellectually hopeless. It could not give rise to reality as we know it. It could not give rise to anything at all. It has no animating desire or intelligence. It is all 'accidental machine' (an oxymoron) with no *Soul* in the machine. If it could account for anything, it could only be for *half* of reality, for matter and mechanism. It has *no consciousness dimensions.* Yet we are conscious and full of emotions, thoughts and ideas. So how could such a mindless mechanism give rise to anything remotely functional, let alone to us?

'Dumb universe or intelligent? Which makes more sense? We are intelligent. So, could not our intelligences be fundamental properties of reality refracted through us? Could not existence itself be intelligent? Could not all mechanisms and forms, ultimately, be *intelligent and ideational* in origin, as the great classical philosophers, like Pythagoras and Socrates, Plato and Aristotle, intuited? Even the belief system called *materialism* consists in *immaterial ideas* about reality. Ironically it does not consist in matter at all. In any case, as Max Planck, the discoverer of quantum physics, explained,

> "THERE IS NO MATTER as such. All matter originates and exists only by virtue of *a force* ... We must assume behind this force the existence of ... intelligent mind. ... mind is the matrix of all matter."[151]

A view echoed by the eminent physicist Sir James Jeans who commented,

> '. . . the Universe begins to look more like a great thought than like a great machine. Mind no longer appears to be an accidental intruder into the realm of matter . . . we ought rather hail it as the creator and governor of the realm of matter.[152]

'Without pre-existing, creative mind, *which just mysteriously Is,* there could *be* no existence. There can be no object, of any kind, without a *subject* to know it. Equally, it takes *consciousness* to imagine a universe without consciousness.

'We cannot get rid of the subjective and the mental. There cannot *be* existence without mind. It's impossible. It takes *awareness* to know there is something rather than nothing. Without *knowing being to know Being* there cannot *be Being,* there can only be oblivion. As Peter Wilburg writes,

> 'the truth is that it is no mysterious 'quantum void' but rather consciousness itself which is the sole reality behind and within all things – and the source of them all. Not 'my' consciousness or 'yours' but consciousness as such, understood as an infinite, all pervasive field of pure [creative] awareness, one latent with infinite potentialities of expression that are constantly manifesting as every existing thing we experience – from a rock to a laptop. For as a saying of the great 10th century Indian sage Abhinavagupta

---

[151] *Das Wesen der Materie* [The Nature of Matter], speech at Florence, Italy (1944) (from Archiv zur Geschichte der Max-Planck-Gesellschaft, Abt. Va, Rep. 11 Planck, Nr. 1797) Emphasis added.

[152] J. Jeans, 1931. *The Mysterious Universe.* Cambridge University press, p. 137.

expressed it so well: "the being of all things that are recognized in awareness in turn depends on awareness."'[153]

'Isn't that strange?' Mused Alisha.

'Well, existence *is* strange. The very *is-ness* of things and our very capacity to *know* are deep mysteries. Existence *is* mysterious, Ali. We can't get away from that. The entire universe is being created and recreated all the time. Creation is not some *one off* thing. It is happening *this* moment and *every* moment. This is why, in his book, *The Science Delusion,* Peter Wilberg writes,

> 'According to quantum physicists such as Hawking, while it may seem that your television or computer came from a factory in China, in reality its every atom and particle is constantly manifesting from an invisible but all pervasive quantum 'vacuum'.
>
> In other words, the mobile phone you pick up or the television or computer screen you look at today is not the one built in a factory–or even the one you picked up or perceived a minute or even a nanosecond ago! For matter – all matter – is constantly emerging from and disappearing back into a quantum vacuum. What exactly this quantum 'field' or 'vacuum' is no physicist can say. Yet what they are saying could and has been said in a quite different way and from a quite different perspective. It was said by those ancient sages, [the yogi-adepts of ancient India] who declared that every aspect of our experienced world – and everything in it – is constantly and continuously manifesting from a vast and infinite field of consciousness.'[154]

'India's yogi-sages, like Abhinavagupta, pointed out, centuries ago, that physical reality is a mysterious Cheshire Cat's SmiLe of a CoSmIc iLluSiOn, made, ultimately, of nothing but S...p...a...c...e..., L...o...v...e..., L...i...g...h...t..., W...a...v...e..., F...o...r...c...e..., F...r...e...q...u...e...n...c...y a n d I...n...f...o...r...m...a...t...i...o...n, or, as Plato said, I...d...E...a...S.... so, ultimately, it's not truly 'solid' or m....a....t....e....r....i....a...l... at all.

'Materialists say, 'Only **matter** is real, so life *has* to be an accident.' Yet, ironically, modern physics shows us that, just as the yogis always said, matter is not primary, Mind and Consciousness are. In deep reality, our world, which seems so solid to us, is not, ultimately, s....o....l....i....d at all. It is made of *shimmering living fields of intelligently in-form-ed energy-space.* If Mind and Spirit are primary, an intelligently guided creation and evolution is possible.'

[153] Wilberg, P. 2008. *The Science Delusion: Why God is Real and 'Science' is Religious Myth.* New Gnosis Publications. P. 141; emphasis and text in square brackets added.

[154] Ibid

17 Unintelligent Design: Is It Possible?

'Really, Ali, we are looking at just two or three fundamental questions here. Firstly, can pure luck create anything functional? Is there any evidence that it can? No, none. Secondly, is there any evidence to show that, despite this, Life *did* start by pure luck? The answer again is a clear, "no."

'Materialism's arguments for the, supposedly, so very amazing creative capabilities of pure stupid luck are based on a refusal to allow that global reality may be *multi-dimensional* and a place, not only of mindless atomic billiard ball matter, but, also, of *mysterious but incredibly creative, Spirit.*

'As Stephen Meyer explains, in *Signature in the Cell: DNA and the Evidence for Intelligent Design,* one thing scientists are supposed to do is to make inferences to the most likely explanation for something. Is pure chance more likely? Is blind chance *'causally adequate'* to create a given mechanism, or is intelligence, even if we cannot directly see its ultimate sources or source, the more logical explanation?

'In *Darwin's Black Box,* biochemist Michael Behe gives many examples where mere chance would not work to assemble living things because the bio-functionalities involved are not just complex, they are *irreducibly complex.'*

'Meaning?'

'Irreducibly complex things have to *arise all at once* or not at all. They cannot arise in an extremely slow, step-by-step process of progressively accumulating complexity as Darwin believed might be possible.'

'Give me an example.'

'We have already met several. The Cell Wall CODED for by the very DNA *within* the cell which the cell's wall protects from *without.* Neither could arise without the other. Avian lungs: different from those of all other vertebrates. They could not arise gradually without causing death – *instantly.*

'The *bacterial flagellum*: the tiny tail bacteria propel themselves with. This fascinating nano-scale tail is powered by a microscopic motor which, at very high magnification, looks remarkably like a tiny electric motor. Look it up, you'll see how incredibly machinelike it is. It does look not organic at all.

'In *Darwin's Black Box,* Michael Behe explains that this intriguing molecular machine could *never* arise by a series of random, Darwinian-style processes, because, just like the motors we make, *all* the working parts would need to be there at the time of assembly or it would not work at all.'

'So only some kind of *subtle, underlying intelligence* could have given rise to the bacterial propeller?' Wondered Alisha.

'Yes. This is the logical conclusion. One way to show intelligent causation – versus unintelligent design in nature – is to demonstrate that a mechanism really could not arise by mere chance. If so, a valid alternative hypothesis is intelligence, which can be inferred, even if we cannot physically see its source.

'Behe explains that there would be no adaptive advantage to just one or two of the flagellum's parts being in place, so the relevant parts would never be naturally selected. The *regular laws of chemistry* would not do it because they cannot give rise to the *irregular but highly specific* ordering of parts needed.

'It makes sense, then, to hypothesize that, just like the motors we make, it's a product of intelligence, of subtle, non-physical intelligence acting within physical reality.

'As an example of irreducible complexity from the mechanisms we make, Professor Behe cites a simple mousetrap, explaining that not only is the trap purposefully ordered, it is all or nothing. It has to be assembled in one go.'

'Because without all its parts there from the beginning it will not work?'

'Exactly. People have attempted to rebut Michael Behe by arguing that 'this or that part of the tiny bacterial motor *might* have been part of something else and it *might* have become a part for the flagellum, and this disproves Dr Behe's theory.'

"Might,' though, is not evidence. It's little better than saying, as some have, 'Look, at one time, this particular part of the bacterial flagellum, as part of another biological system, *might* have luckily combined with other parts to become the bacterial flagellum. This shows Professor Behe is wrong. The little motor is not irreducibly complex.'

'This, however, begs the question of where that other part came from in the first place and it is about as useful as claiming that,

> 'Aircraft are unintelligently designed. How do we know? Because they have propellers and so do boats. So, somehow, at some point in evolutionary history, a boat propeller *might, by chance,* have become attached to an airplane. This shows that aircraft are unintelligently designed.'

'Michael Behe is right. It is easy to say, "Oh, this or that thing *"just evolved,""* without bothering to think very much about what would really need to be involved. This is intellectually sloppy and it's intellectually lazy.

'In his book, *No Free Lunch: Why Specified Complexity Cannot Be Purchased without Intelligence,* William Dembski, who is neither sloppy nor lazy, shows how unguided, Darwinian-style mechanisms are incapable of generating the special kind of complexity which describes the wholly specific, the entirely purposeful, and often extremely long sequences of information and parts which define all biological mechanisms.

'To try to negate such claims, some contemporary Darwinians, including Richard Dawkins, have *intelligently designed* computer programs to try to show that living mechanisms could, in fact, evolve by random, no-intelligence-needed processes. Dembski shows, however, that all such computer generated scenarios fail because *they smuggle in pre-defined search criteria* which are neither random, nor unintelligent, nor available in the directionless and 'purposeless' nature of materialist belief, while deluding themselves and their audiences that they do not. A reviewer of *No Free Lunch* writes,

> 'Dembski describes the 'shell game' [three cups trick] whereby evolutionary programmers try to obscure the 'smuggled in' information and claim that their programs have generated the output[ted] information from scratch. ...
>
> 'This book marvelously extends and expands Dembski's earlier work. I highly recommend it. My only hope is that critics will have the decency to actually engage the ideas presented here. ... Dembski has ... 'taken on' the established Darwinian orthodoxy, demonstrating the insufficiency of non-intelligent processes to generate [the] specified complexity [that defines all Living things].' [155]

'Dembski is correct. Neither regular physical laws, nor random processes, are capable of generating the purposeful – yet irregular – sequences of coding information and parts which define all the living mechanisms Behe discusses.

'If you read his account, in *Darwin's Black Box,* of the *extreme* complexities of the biochemistries of vision and of the blood clotting cascade, you will see how unrealistic and how shockingly unscientific it is to attribute such amazingly complex processes to mere luck.

'Dr Behe's arguments for intelligent design make sense. He's not making it up. He shows it's *simply not possible* for systems like the bacterial flagellum to arise from the accumulation of small, random steps. It's only if we have a prior commitment to materialism that we will be forced to deny this conclusion. A damaging commitment which blocks perfectly reasonable lines of inquiry.

'In *Billions of Missing Links: A Rational Look at the Mysteries Evolution Can't Explain,* G. S. Simmons also gives many examples of biological mechanisms which cannot, reasonably, be explained in purely chance terms.

'For example the *irreducibly complex* defense technologies of insects:

---

[155] Dembski, W. 2007. *No Free Lunch: Why Specified Complexity Cannot be Purchased without Intelligence.* Rowman Littlefield. 'Very Insightful,' 4. 22.02, a customer, amazon, text in square brackets added.

'An Australian termite called the nasui ... If provoked by a predator, it will swing its head side to side and squirt loops of a chemical mix. One component will glue the attacker in place, another will paralyze it, another may kill it, and a fourth will attract other termite soldiers to gather around the intruder and to keep spraying it. Their sprays contain alpha-pinene, beta-pinene, limonene, trinervitrenes, and kempanes. The latter two chemicals cause the stickiness and *they have not been found anywhere else in nature.* This species also defies evolutionary logic.'[156]

'Because?' Asked Alisha.

'Because no such highly specific and purposeful combinations of chemicals or behaviors could *ever* arise by chance, no matter how much time you allow.

'As Dembski has explained, the statistical odds against it are insurmountable, utterly insurmountable. Simmons describes another fascinating example of irreducible complexity, the bombardier beetle's weaponry.

'This African insect can fire off two chemicals, hydrogen peroxide and hydro quinone, from *separate storage tanks* and rear jets. [The beetle can fire accurately and repeatedly. While separate the compounds are harmless. When mixed they instantly heat up and hit their opponent at boiling point, 100°C!] When the chemicals combine, they form a new chemical that burns the predator. The beetle can shoot these chemicals with an uncanny accuracy, as well, to either side, backward, or even forward, by swinging its tail under its abdomen. Special nozzles blast predators at a rate of 500 bursts per second, each at a speed of 65 ft./s. These chemicals are potent enough to severely damage a mouse and injure the eyes of any animal . . . Yet these chemicals are entirely benign when stored *separately* at the back end of these beetles. How could this happen by accident? "Oops, those two chemicals didn't work" (spoken by an intermediate species). "Mind if I try two others before you eat me?" Or, "Could you stand a little taller so I can get you with my nozzles?"

'Keep in mind there are hundreds of thousands of chemicals on this planet to choose from. And even if the combo turned out perfectly right the first time, the beetles still needed a way to make them, store them, and fire them off. [157]

---

[156] Simmons, G. 2007. *Billions of Missing Links.* Harvest House; pps. 207–211. Emphasis added.

[157] Ibid, p. 29. emphasis and text in square brackets added.

'Are these all merely lucky chemical mutations [genetic copying errors] that protected the species? ... How did the special storage glands that protect the owner evolve? Before or after? All at the same time?

What happened to the thousands of missing links? [Where is the fossil or other evidence for them?]' [158]

'As Simmons says of the Bombardier's fabulous defense system,

'The entire mechanism defies evolution. How and why would there have been a gradual development of chemicals that were useless until they finally reached their present form? This system is all or nothing [it's irreducibly complex].'

'Another example of irreducible complexity comes from the most dynamic level of all, the *intracellular* level. Living Cells, like vast, chemical-factory-worlds, in spaces smaller than sand grains *(think about that!)*, contain nano-scale bio-molecular-robots called proteasomes.

'An article in *Nature* magazine, discussing these, begins: 'Correctly dismantling a structure can be as challenging as assembling it. The architecture of the yeast proteasome reveals this enzyme's intricate machinery for protein degradation.' [159] Commenting on this article, *Evolution News* wrote:

'Thus begins an article in *Nature* about the cell's shredder-recycler, the proteasome. This large, barrel-shaped molecular machine ... validates the trash, pulls it in with a motor, and shreds it inside. . . . Inside the barrel . . . active sites on the inside walls cleave polypeptide chains into short segments about 7–9 amino acid units long, which can be reused by the cell directly, or further degraded into individual amino acids for recycling. . . . An elaborate lid structure, composed of 19 more protein parts, guards the entry gate and checks the credentials of each protein that enters.

To be validated, a target protein must have been previously tagged for destruction by other molecular machines. . . . Once tagged, the doomed protein is conveyed to the proteasome. . .

Proteasomes move [24/7] throughout the cytoplasm and the nucleus carrying on their vital work every day inside your body, identifying and selecting proteins to degrade, keeping the Cell free of damaged or mis-folded proteins, and removing proteins whose work is done. ... These intricate *machines* help keep the immune system functioning and are ever-present to

---

[158] Ibid, pps. 207–211,

[159] Geng Tian and Daniel Finley, "Cell biology: Destruction deconstructed," *Nature* 482, (09 February 2012), pp. 170–171, emphasis added.

respond to stress. Failure of these systems can cause serious diseases, like Parkinson's and Alzheimer's, or death.[160]

'The mainstream view of this incredibly sophisticated functionality, Ali, is that *'somehow, it must have evolved. . . by chance!'* *Evolution News* notes that although the 'authors of the original paper said nothing about evolution,'

'Geng Tian and Daniel Finley, though, in their summary of the paper in *Nature,* couldn't resist sprinkling a little Darwin-brand sneeze powder on stage. "The eukaryotic proteasome seems to have *evolved* from a protease known as PAN (or something comparable to this enzyme), which is found in microorganisms called archaea," they suggested. Then they produced a colorful diagram called *"The evolution of* proteases," showing trypsin, PAN, and the full-fledged proteasome, as if to mimic those outworn icons of evolution, the horse series and [the] monkey-to-man parade.

If all else fails, say it with feeling: "This dramatic *evolutionary elaboration* of the protein-degrading machinery is reflected in the fact that the proteasome has assumed regulatory functions in virtually all aspects of eukaryotic cell biology" ... *"The evolution of* ubiquitin tagging also coincided with a transformation of the proteasome's structure".

[However] *did they show how mutations [random genetic copying mistakes] and unguided processes produced this highly complex machine,* let alone the other systems it interacts with? No; they just assumed it. *"The evolution of"* begs the question. Comparing three functional designs of varying complexity doesn't prove they evolved any more than showing a scissors, a shredder and a recycling system proves they evolved from each other by *undirected* processes.'[161]

'They conclude, '*"The evolution of"* is a useless if ubiquitous phrase that needs to be tossed into the recycler.' Remember, as biochemist, Michael Behe, wrote,

'There is no publication in the scientific literature in prestigious journals, specialty journals, or books that describes how molecular evolution of any real, complex, biochemical system either did occur or even might have occurred. There are [many breezy, lazy,] *assertions* that such evolution occurred [as in the article above], but *absolutely none are supported by pertinent experiments or calculations.'*[162]

---

[160] *Evolution News,* 16th February, 2012; emphasis and text in square brackets added.

[161] Emphasis and text in square brackets added. References cited in the Discovery article: (1) Geng Tian and Daniel Finley, 'Cell biology: Destruction deconstructed', *Nature* 482, 9 February 2012, pp. 170–171. (2) Lander *et al.,* 'Complete subunit architecture of the proteasome regulatory particle', *Nature* 482, (9 February 2012), pp. 186–191.

[162] Behe, M. 1996. *Darwin's Black Box.* The Free Press, p. 185. Emphasis and text in brackets added.

'It is strange, Ali, how the naive, bible-based, medieval creationism, where no one doubted that Spirit creates everything, has been replaced by an equally simple kind of *faith* that, "Oh, just give it enough time and, "whatever you like," no matter how complex, will "evolve," no intelligence or purpose need ever be involved." Don't people see how intellectually vacuous and downright lazy this is? With no attempt at all to explain precisely *how* aimless, mindless, chemical chance could create anything remotely clever?

'You know something, Ali, a lot of people, today, don't realize that what we think of as *Darwin's Theory* actually had two inventors and that it was, initially, thought of as belonging to both of them, not just to Darwin.

'Who was the other one?' Asked Alisha, looking intrigued.

'Alfred Russell Wallace. Initially, his ideas were very similar to Darwin's. But he, unlike Darwin, eventually realized that no *wholly materialistic* theory – be it theirs or anyone else's – could provide a full explanation of Life.

'Interestingly, he also investigated the subject of *spiritualism* which was very popular at the time and, along with other luminaries, like Sir Arthur Conan Doyle (literary genius), Sir Oliver Lodge F.R.S. (world class scientist and developer of spark plugs), Professor James Mapes (expert in chemistry), Frederick Myers (classicist and dedicated psychic researcher), Sir William Barrett F.R.S (professor of physics), Professor Camille Flammarion (founder of the French Astronomical Society), he concluded,

'that the phenomena of Spiritualism in their entirety *do not* require further confirmation. They are proved quite as well as facts are proved in other sciences.' [163]

'But, apart from the evidences from psychic research, there were other things, Wallace realized, which no materialist theory of Life could explain.

'Like?'

'Like the extraordinary metamorphosis of the crawling caterpillar into the beautiful Butterfly.[164] Wallace realized there was *no* way mere chance could explain the incredibly complex bio-chemical transformations involved. The astonishing Caterpillar to Butterfly metamorphosis shows very well the causal inadequacy of *both* Darwinism and materialism to explain living things.

'A caterpillar, Wallace realized, is a perfectly viable organism in its own right. There's **no reason whatsoever** for the random caterpillar of Darwinian

---

[163] Quoted by Michael E. Tymn, http://whitecrowbooks.com/michaeltymn/biography/ "It wasn't long after the birth of modern Spiritualism in 1848 that scientists and scholars began investigating the phenomena. Many of them started out with the intent of showing that all mediums were charlatans, but one by one they came to believe in the reality of mediumship and related psychic phenomena." Tymn,

[164] M.A. Flannery, 2011. *Alfred Russell Wallace: A Rediscovered Life,* Discovery Institute Press

thinking to randomly dissolve itself into a purposeless and life-forceless[165] chemical soup, then to 'accidentally' re-emerge as a gorgeous, *living*, Butterfly.

'It is *scientifically absurd* to claim that a rare genetic copying mistake would cause a caterpillar to 'mutate a little' towards making an accidental cocoon. As molecular biologist Michael Denton writes,

> 'The design of living systems, from an organismic level down to the level of an individual protein, is *so integrated* that most attempts to engineer even a relatively minor functional change are bound to necessitate *a host of subtle compensatory changes*. It is hard to envisage a reality *less* amenable to Darwinian change via a succession of independent undirected mutations altering one component of the organism at a time.[166]

'Denton is right. Even if a caterpillar accidentally started to make a cocoon, what would happen? Let's imagine it so *we* can't be accused of being intellectually lazy. Ok, so one Caterpillar member of the original species randomly mutates and, by pure chance, it becomes a little cocoon-like. Although, why would it? After all, all cells *actively protect against* random genetic changes, by extremely sophisticated means.

'Then, even if such an unlikely thing ever happened, *why* would such a change be naturally selected? It would confer no survival advantage, nor lead to constructive change, let alone to a beautiful Butterfly, but to *death!*

'It would be lethal. There'd be absolutely nothing to cause the ex-caterpillar to return to being a caterpillar or forward to being a never-before-seen butterfly. And if such an occurrence happened more than once, for all caterpillars starting to mutate into a cocoon, *only death would await*. There is no selective advantage in merely *starting* to form a cocoon. What, after all, would ever cause the, 'it-used-to-be-a-caterpillar-now-it's-an-accidental-soup-in-a-cocoon' to randomly morph into a beautiful Butterfly involving literally *millions* of highly integrated and incredibly complex biochemical changes, not just one or two? Can commonsense no longer play any part in science when it comes to our inquiries into the genesis of living things?

'What random Darwinian process, for heaven's sake, would cause the 'accidental' and 'purposeless'[167] goo the Caterpillar had dissolved itself into, to turn into a Butterfly? It is scientifically illogical – actually it's absurd – to suggest that some unintelligent, Darwinian-style process of rare, generally

---

[165] Since the artificial synthesis of urea in 1828 it is the 'scientific' position that there are no Life Forces.

[166] Denton, M. 1998. *Nature's Destiny*, p. 342, emphasis added.

[167] According to modern science all things in Nature are just chemical 'coincidences' and 'purposeless.'

harmful, genetic copying errors would give rise to the miraculous *bio-logical* happening that is a Butterfly. Bio = Life! Logos = the Word!, the Soul of IDEAS OR INFORMATION. It's interesting to note that to 'in-form' and 'in-form-ation' can be understood as to 'bring into form.'

'Darwin observed that skillful breeders could modify various species, like pigeons, cattle and dogs, to *some, ultimately, limited, species-bound* degree. Fine. This is uncontroversial micro-evolution. But to extrapolate from this, as he did, that one species could change into an entirely other kind of species is unjustified by the data. Yet, this was his *core,* macro-evolutionary idea.

'It is true that Darwin and, initially, Wallace not having our modern knowledge, did not realize that Life could *never* arise by 'chance,' nor *evolve* by such means, that this was a statistical and chemical impossibility.[168] Had he known these things, Darwin would, surely, have conceded that,

> 'The intriguing idea that Life might just be an accident which evolved by accident is, it is now clear, *impossible.* It has been demonstrated repeatedly that *neither living cells,* nor such transformations as the fabulously complex caterpillar-to-butterfly process could, as I once believed, 'have been formed by numerous, successive, slight modifications.' This means that, as I, myself, once said, 'my theory would absolutely break down.'
>
> 'To attempt to explain Life in purely materialistic terms – as a long series of extremely improbable chemical coincidences – no longer makes sense.
>
> 'In light of what we now know, it's obvious that there's an amazing second-by-second, *a 24/7, intelligence* to the functioning of Living Things. It's irrational to deny it.
>
> 'Intelligent causation and the supernatural, however mysterious to us they may seem, are part of the *basic fabric* of reality. If we are to make any further progress in the intellectual West we must accept this, even if, to some of us, it feels like an intellectual defeat and a blow to the pride of our prior faith and beliefs; our faith in pure chance as a supreme creative principle.'

'Darwin wrote, in *The Origin of Species,*

> 'If it could be demonstrated that any complex organ existed, which could not possibly have been formed by numerous, successive, slight modifications, *my theory would absolutely break down.* But I can find out no such case.'

'In light of our knowledge today, Ali, it would make as much sense for contemporary scientists, technicians and engineers to proclaim that,

---

[168] Dembski, W and J. Witt. 2010. *Intelligent Design Uncensored.* IVP, p. 72.

'If it could be demonstrated that any complex, AI guided Robot, Car, TV or Computer, [all far simpler than any living thing], could not possibly have been formed by numerous, successive, slight modifications, driven by nothing but the [unguided and allegedly 'purposeless'] laws of physics in combination with numerous lucky, 'natural' events [random swirling about of the parts] our theory of evolution by random – unintelligent – causation would absolutely break down. But we can find out no such case.'

'Such a statement would be absurd. But when it comes to living things the usual rules no longer seem to apply. There is *nothing* in the unguided laws of chemistry to cause *any* living thing to arise. It is *easy* to show that Life could *neither arise nor evolve* by such means; even some materialists have pointed this out. And there is *no* good evidence to support Darwin's overall scheme.

'For it's time, Darwin's idea was interesting, although it was very much driven by his materialist philosophy and of jumping to conclusions far too quickly. This was precisely *why* various distinguished scientists of the day, and since, pointed out key flaws in his theory from the start.

'But, for the fashionable intellectuals of his time, Darwin's ideas were a convenient way to sideline the traditional religious authorities and to make materialism and secular thinking, not classical idealism and spiritualism, arbiters of the highest truths.

'Darwin believed that Life could arise due to random, unguided, chemical chance. He imagined that an original, single-cell microbe might first arise by chance, then, progressively, morph into a multi-celled worm or trilobite or spider, that an invertebrate, like a worm or trilobite, 'might' gradually morph into a vertebrate fish, a fish into an amphibian, amphibian into dinosaur, dinosaur into bird and mammal, and, finally, mammals into people.

'Because he was so attached to his theory, he chose to ignore the *clanging* lack of fossil evidence for his ideas. Equally, he disregarded the accumulated evidence of the world's most skillful breeders that it is not possible, nor has it ever been possible, to change any species beyond certain genetic limits which no breeding program, no matter how intensive, has ever been able to cross. But we, who now know so much more than he, cannot continue in this way.

'Today, we have, not just one or two but many demonstrations of the kind which Darwin said could be used to show that his theory of macro evolution by random variations away from a hypothetical universal common ancestor microbe, "would absolutely break down." They include, among others,

1. Life [169] and Living Cells the stunning complexities of which Darwin and his steam era contemporaries knew very little.

2. Irreducible Mind. Mind which cannot be reduced to *mindless* matter.[170]

3. The *irreducibly complex* bacterial flagellum and its nano-scale motor, [171] of which Darwin and his 19th century contemporaries knew nothing.

4. The *irreducibly complex* blood clotting cascade and the *irreducibly complex* biochemistry of vision,[172] of which Darwin knew nothing.

5. The *irreducibly complex* defense technologies of insects discussed by Simmons in *Billions of Missing Links.*

6. The metamorphosis of the Butterfly considered by Alfred Russell Wallace.

7. Life's complex digital (DNA) CODES which, like all codes, could never arise by mere luck, and, of which Darwin knew nothing.

8. Life's thousands of *specifically ordered* protein sequences, which could never arise by chance, and, of which Darwin knew nothing.

9. The arising of Bird's Feathers and *irreducibly complex* Avian Lungs discussed by Michael Denton, neither of which could possibly arise due to relatively rare, random genetic mutations – genetic copying errors.

10. The evidence from the fossils which *refutes* Darwin's ideas.

11. The problem of *sudden emergence which* refutes Darwin's ideas.

12. The tremendous *stability* of the species which *refutes* Darwin's ideas.

13. The evidence from embryology which *refutes* Darwin's ideas.

14. The evidence from genetics which *refutes* Darwin's scheme.

'In view of all these things, Ali, don't you think it's time for our sciences to have a serious rethink about origins?'

We walked on.

---

[169] George Wald, 'The Origin of Life', *Scientific American* 191:48 (May 1954). Eastman, M. C. Missler. 1995. *The Creator Beyond Time and Space*

[170] Kelly, E. F. E.W. Kelly, A. Crabtree, A. Gauld 2009 *Irreducible Mind: Toward a Psychology for the 21st Century.*

[171] Behe, M. 1996. *Darwin's Black Box.* The Free Press.

[172] Ibid.

18 Mind Over Matter: Intelligent Design

'Unfortunately, Ali, too many people today, are not encouraged to think for themselves. Their teachers, who were subjected to the same teachings, tell them that Darwinian macroevolution is a no brainer while making fun of those who look a little deeper and wonder if this is really true.

'So many have been led to think that they are just 'purposeless' coincidences of the *lifeless rocks and mindless liquids* of our planet. Really? Should not this extraordinary claim be backed up by some pretty extraordinary evidence before we accept it?

'For example, how many ever truly contemplate our thinking and feeling and sensing? The *mindless little atoms,* of which our physical bodies are made, have no reason or capacity, within the materialist scheme, *to think or to feel, or to touch, taste, see, smell and hear, – do they?*

'Yet we can think and feel and we have amazing senses. So how can it be 'scientific' to assert that our capacities to *feel* and to *know* must just be lucky coincidences of insensate atoms of carbon or hydrogen?

'Or, think of your joys or your worries. Is the insentient matter of the materialist belief system capable of worry or joy? No! It is insensible, dead! Without mind or feeling! It is something *else* in us, then, not the dead matter of materialism, which can *know and love and desire,* which can think and create and feel joy or pain. Classically, people understood that it is the *Soul* in us which can sense and look and touch and know. But, according to our present science, we do not have *Souls nor, even, real Minds.*

'Neither Darwinism nor materialism can account for Life or Mind. Their adherents seem unable or unwilling to understand that no amount of examining brain chemicals will ever locate the content of our *immaterial thoughts or our immaterial ideas* anymore than analyzing the chemicals from which the parts of a TV (the body) are made will ever find the Show (the Soul) which is being transmitted at the time.

'The distinguished philosopher, Thomas Nagel writes,

'The existence of consciousness is... one of the most astounding things . . . if physical science, whatever it may have to say about the origin of life, leaves us necessarily in the dark about consciousness, that shows that it cannot provide the basic form of intelligibility for this world. There must be a very different way in which things as they are make sense. . .[173]

'He adds,

'For a long time I have found the materialist account of how we and our fellow organisms came to exist hard to believe, including the standard version of how the evolutionary process works. The more details we learn about the chemical basis of life and the intricacy of the genetic code, the more **unbelievable** the standard historical account becomes.. [and] it seems to me that.. the current orthodoxy about the cosmic order is the product of governing assumptions that are unsupported, and that **it flies in the face of common sense**.'[174]

'In *Irreducible Mind: Toward a Psychology for the 21st Century* the authors write,

'Current mainstream opinion in psychology, neuroscience, and philosophy of mind holds that all aspects of human mind and consciousness are generated by physical processes occurring in brains. ... this reductive materialism is *not only incomplete but false.*

'The authors marshal evidence for a variety of psychological phenomena that are extremely difficult, and in some cases clearly impossible, to account for in conventional physicalist terms: memory, psychological automatisms and secondary personality, near-death experiences and allied phenomena, genius-level creativity, and 'mystical' states of consciousness . . .'[175]

'Raymond Tallis, a neuroscientist, also fiercely criticizes the materialist attempt to reduce people to their physical brains. In *Aping Mankind: Neuromania, Darwinitis and the Misrepresentation of Humanity,* he blasts,

'the exaggerated claims made for the ability of neuroscience and evolutionary theory to explain human consciousness, behavior, culture and society. ... Tallis directs his guns at neuroscience's dark companion – Neuromania, as he describes it . . . Tallis dismantles the idea that "we are our brains", which has given rise to a plethora of neuro-prefixed pseudo-

---

[173] Nagel, T. 2012. *Mind and Cosmos: Why the Materialist Neo-Darwinian Conception of Nature is Almost Certainly False.* Oxford University Press.

[174] Ibid, emphasis added.

[175] Kelly, E. F. E.W. Kelly, A. Crabtree, A. Gauld 2009 *Irreducible Mind: Toward a Psychology for the 21st Century.* Rowman and Littlefield, product description, emphasis added.

disciplines ... shows it to be confused and fallacious, and an abuse of the prestige of science, one that sidesteps a whole range of mind-body problems. . . . offering a grotesquely simplified and degrading account of humanity.'[176]

'Peter Wilberg, in *The Science Delusion,* is also blunt. He writes,

'. . . the rationality of scientific thinking as such collapses in on itself as soon as we accept the current 'scientific' belief that 'rational' thinking as such is merely the autonomous, unwilled biological activity of some bodily 'thing' – *the brain* – rather than a free and independent activity of *mind.* For how, and by what independent 'rational' criteria, can its 'rationality' be judged? [This kind of] Science then, does not lead us up into the heights of free and rational thinking, but is instead its very nemesis – an irrational, self-contradictory world-view parading itself [like a naked emperor] as the very apotheosis of rationality.'[177]

'In *Science and the Near Death Experience,* Chris Carter notes,

'The materialist believes that consciousness is created by matter, yet the best theory we have about the nature of matter [quantum physics] seems to require that *consciousness exists independently of matter.* And materialist models of mind utterly fail to answer the hard problem: why should consciousness exist in the first place and then constantly deceive us as to its function?'[178]

'From the materialist belief that existence lacks any inherent meaning or purpose must flow the claim that Life and Mind are cosmic accidents. That this doesn't work is ignored. Why? For emotional and philosophical reasons. As the philosopher John Searle writes,

'Acceptance of the current views is motivated *not* so much by an independent conviction of the truth as by a terror of what are apparently the only alternatives. That is, the choice we are tacitly presented with is between a 'scientific' approach, as represented by one or another of the current

---

[176] Tallis, R. 2011. *Aping Mankind: Neuromania, Darwinitis and the Misrepresentation of Humanity.* Acumen Publishing; product description.

[177] Wilberg, P. 2008. *The Science Delusion: Why God is Real and 'Science' is Religious Myth.* New Gnosis Publications. P. 123–124; text in square brackets added.

[178] Carter, C. 2010. *Science and the Near Death Experience.* Inner Traditions; p. 74. emphasis added.

versions of 'materialism,' and an 'anti-scientific' approach, as represented by Cartesianism or some other traditional religious conception of the mind.'[179]

'He's right because many people now conflate science with materialism and they consider all ideas of the supernatural to be delusion. Alongside this rejection of classical dualism goes the modern dogma that *spiritual* mind could not influence 'non-spiritual' matter because how could 'unlike' (mind) act on 'like' (matter)? This idea, if true, would mean that an intelligently guided creation or evolution is impossible.'

'So the belief that only 'like' can act on 'like' is key?' Remarked Ali.

'Yes. It's a key tenet of materialism that mind cannot exist apart from matter nor, if apart, that it could influence it in any way. Because it would be a different principle. Yet there are many ways to show this to be false.

'How?'

'Well, there is good quality modern evidence that mind-to-matter communication at a distance *is* possible. For example, it has been repeatedly demonstrated that people can remotely influence random number generators.[180]

'Gifted spiritualist mediums have also shown, in very carefully controlled experiments,[181] that they can transmit verifiable messages from those passed on. And many people have described near death[182] and out of body[183] experiences, which, taken in the round, point towards multi-dimensionality and to mind entirely *beyond* the body.

'So, if our minds can influence random number generators, if we can operate beyond the body and if, after death, we can, sometimes, continue to communicate with those still in the physical, it implies that there may be *levels of non-physical mind,* up to and including spiritual or divine mind?' Mused Alisha.

'Yes, interestingly, on this key point, philosopher of science, Neal Grossman, was asked how could *unlike* (mind) act upon *like* (matter), because

---

[179] Searle, *The Rediscovery of the Mind,* 3–4.

[180] Radin, D. Phd. 1997. *The Conscious Universe: The Scientific Truth of Psychic Phenomena.* Harper Collins.

[181] Schwartz, G.E. Phd. W.L. Simon. 2002. *The Afterlife Experiments: Breakthrough Scientific Evidence of Life After Death.* Atria Books.

[182] Carter, C. 2010. *Science and the Near Death Experience.* Inner Traditions; p.235.

[183] Gustus, S. 2010. *Less Incomplete: A Guide to Experiencing the Human Condition Beyond the Physical Body.* O Books. Monroe, R.A. 1972. *Journeys Out of the Body.* Souvenir Press. Buhlman, W. 2001. *The Secret of the Soul: Using Out-of-Body Experiences to Understand Our True Nature.* Harper One. Peake, A. 2011. *The Out of Body Experience: The History and Science of Astral Travel.* Watkins Publishing.

doesn't science say the universe is a closed system and, therefore, there is no way *in* for spiritual mind to influence it or to create anything? He replied,

'On general principles, it is unscientific to claim that one's current theory is true a priori. *Facts trump theory.* Psychokinesis [mind-to-matter interaction] is a fact that has been unequivocally demonstrated over and over again (e.g. the PEARS experiments). It is absurd to deny that the facts are what they are simply because their occurrence poses difficulties for our current theories. It is not good scientific practice to reject empirical data because it poses problems for conventional theories. *Materialism is clearly false* ... The point is that we must stay close to the data, even if there is no theory that adequately explains it. Anomalous data is the engine that drives the search for new theories, even though, as the history of science shows, *there will always be those who prefer their old theories to new data that falsifies what they believe.* So it is a fact that mind can influence a random number generator; we do not now have a theory that can explain this fact, and it is highly unlikely that such a theory will be compatible with Materialism.'[184]

'Grossman is right. Facts (should) trump theory. We've seen this with Darwin's ideas where the fossil and other facts falsify the theory, but the theory persists, not because it's well proven, but for reasons of habit and belief.

In his fascinating book, *Science and the Near Death Experience,* Chris Carter observes that materialists consider Cartesian dualism, (Mind *vs* Matter, Soul *vs* Mechanism, Life *vs* Form thinking), to be outdated and pre-scientific. But not all scientists think this way. The astronomer V. A. Firsoff wrote,

'to assert there is *only* matter and no mind is the most illogical of propositions, quite apart from the findings of modern physics, which show that there *is* no matter in the traditional meaning of the term.'[185]

'Chris Carter also notes the irony that while biologists have been trying to bring biology into line with the reductive materialism typical of 19th-century physics, where, as Francis Crick put it,

'The ultimate aim of the modern movement in biology is to explain all biology in terms of physics and chemistry,' [186]

---

[184] www.subversivethinking.blogspot.co.uk/2011/05/interview-with-philosopher- neal.html Emphasis and text in square brackets added.

[185] Carter, C. 2010. *Science and the Near Death Experience.* Inner Traditions – Quoted in Koestler, *The Roots of Coincidence,* 77.

[186] Crick F. 2004. *Of Molecules and Men.* Prometheus, p. 10.

'At the same time,' Harold Morowitz, professor of molecular biophysics at Yale, writes,

'physicists, faced with compelling experimental evidence, have been moving away from strictly mechanical models of the universe to a view that sees the mind as playing an integral role in all physical events.' [187]

'Carter adds,

'... we have known since the early years of the 20th century that classical physics fails drastically at the atomic and subatomic levels, and that the behavior of such particles is deterministic and *observer dependent*. The irony here is that while materialists often describe themselves as promoting a scientific outlook, ***it is possible to be a materialist only by ignoring the most successful scientific theory of matter the world has yet seen***. The materialist believes that consciousness is created by matter, yet the best theory we have about the nature of matter [quantum physics] seems to require that consciousness exists independently of matter.' [188]

'Carter also tackles the point put to Grossman that causal interaction between Mind and Body, the classic *dualistic* hypothesis, conflicts with,

'the physical theory that the universe is a *closed* system and that every physical event is linked with an antecedent physical event. This assumption preempts any possibility that a mental act can cause a physical event.'[189]

"Of course,' Carter continues, 'we know now that the universe is *not* a closed system and that the collapse of the wave function – *a physical event* – is linked with an antecedent *mental event*. The objection K. R. Rao describes is of course based on classical physics.'[190] Carter quotes Rosenblum and Kuttner who write:

'Some theorists deny the possibility of *duality* [mind-to-matter-causation] by arguing that the *signal from a nonmaterial mind* could not carry energy and thus could not influence material brain cells. ... However, no energy need be involved in determining to which particular situation a wave function collapses. . . . *Quantum mechanics therefore allows an escape from the*

---

[187] Morowitz, "Rediscovering the Mind," 12. Emphasis added.

[188] Carter, C. 2010. *Science and the Near Death Experience.* Inner Traditions; p. 74. emphasis added.

[189] Rao, "Consciousness."

[190] Carter, C. 2010. *Science and the Near Death Experience.* Inner Traditions; p. 75. emphasis added.

supposed fatal flaw of dualism. It is a mistake to think that dualism [mind–matter interaction] can be ruled out on the basis of physics.' [191]

'The well known philosopher of science, Karl Popper, also rejected the concept that only like can act upon like and noted that this belief rests on outdated ideas about physics. Empirical examples of *unlikes* acting on each other are the interactions between the four known and very different forces of physics: gravity, electro-magnetism and the strong and weak forces.

'So the idea that the universe could not be influenced by mind is false?'

'Yes, and it's counterfactual. We have minds but, according to science today, they are not ontologically real. They do not really exist. This is why the famous materialist philosopher, Daniel Dennett, writes, 'We're all zombies. Nobody is conscious.' [192] But this, Ali, is an *irrational* kind of 'rationalism' where rationality is equated with materialism – even when it leads nowhere.

'In the real world, pure chance cannot assemble a very simple wooden toy. Yet, in order to conform to the prevailing view and to try to make materialism seem 'true,' we are asked to contort ourselves logically, intuitively and experientially, to ignore all the evidence which shows it to be false, to deny every single psychic, mystical or spiritual experience, no matter how convincing, we or anyone else has ever had, to agree to the complete meaninglessness of our existences and to thank Darwin for it.'

'So why don't those who dominate science today give way?'

'Partly because of their belief that *spiritual mind* cannot influence matter, partly because the deeper causes of living things cannot be physically seen, partly due to life's difficulties and partly due to professional pride.'

'Don't they have some points there?' Said Ali.

'Well, everyone accepts that life's deeper causes, if real, are not obvious, that existence can be challenging and that it can be hard to change our views. If it were otherwise there would be far less to discuss. But if we were really made of nothing more than mindless atomic elements, like carbon, we'd know nothing of kindness or cruelty, pain or pleasure let alone care about such things. It's only the *Soul* which knows these things, not the dull little billiard ball atoms of the materialist belief system.

'But Charles Darwin, the once young and ambitious Victorian naturalist, in pursuit of ever lasting scientific glory, not knowing what we now know, wondered whether an organ as complex as an eye or brain, or an entire animal, could arise – from mindless cosmic dust – by pure chance. He wrote,

---

[191] Rosenblum and Kuttner, "Consciousness and Quantum Mechanics: The Connection and Analogies," 248, emphasis and text in square brackets added.

[192] Dennett D. 1992. *Consciousness Explained.* Back Bay Books, p. 406.

'IF it could be demonstrated that any complex organ existed, which could not possibly have been formed by numerous, successive, slight modifications, [unguided by any forces beyond pure chance,] *my theory would absolutely break down*. But I can find out no such case.' [193]

'To which we, who owe a debt not only to the past but a promise to the future, must reply,

'**Dear Mr. Darwin**, when you laid down your famous challenge, you could just as easily have written,

"IF it could be demonstrated that any complex organ existed, which could not possibly have been formed by subtle, *super* natural causes, our classical theory of the intelligent causation of Nature's many complex organs and organisms would break down. But we can find no such case."

'In fact, dear Darwin, many experiments have been carried out, over the years, which have successfully demonstrated that, in conflict with your materialist beliefs, mind-to-matter causation *is* possible. Mind *can* influence matter. This overcomes one of the major materialist objections to all ideas of supernatural causation – instantly creative or gradual and evolutionary.

'At the same time, your theory accords us no idea why even the beginnings of an eye or a light sensitive spot of some kind would arise by pure chance? Why would it?

'It is not enough to assert, *"It must have, because that's what our ideas require."* This is not scientific. In any case, a light sensitive spot needs an *inner Being* to receive its messages. *Without a Subject or Soul or Knower to know,* all sensors are pointless. Yet, according to your materialist philosophy, there *are no Souls or true Subjects, no "I-Am-I" Consciousness, beyond space and time.*

'No, there are just accidental bio-robots (which is an oxymoron) who only *imagine* they exist (counterfactual). Just meaningless coincidences of the dead rocks and lifeless liquids of our planet and, when they die, that's all they ever were.

'Yet, since your day, it has been demonstrated, over and over again, that mere 'chance' cannot assemble even the simplest of working mechanisms, not even a simple wooden toy. The refusal to acknowledge this, on the part of those who still follow you, is not only illogical, it's worryingly unscientific.

'Life's deeper sources are not physically visible, we all agree. But this, by itself, doesn't mean they do not exist. There have been many testaments, many corroborated and some verifiable, to the subtle, supernatural or spiritual dimensions

---

[193] Darwin, C. 1859. *On the Origin of Species.* Emphasis added.

of things. It is unscientific to discount this large body of evidence, much of it modern, not ancient or religious, just because it doesn't match your worldview.

'Since you published your ideas, when so much less was known, it has become clear that not just one but many of Life's mechanisms "could not possibly have been formed by numerous, successive, slight modifications," as you once imagined might be possible. This means that, as you yourself said, your "theory would absolutely break down..." Yours sincerely, etc.

'Since Darwin's day, Ali, it has been shown again and again that many features of Living things could not possibly have been "formed by numerous, successive, slight modifications." Darwin provided *no* evidence to support his grandiose claims for the allegedly so very supreme creative capabilities of pure stupid chance. As Michael Denton writes,

> 'To common sense it seems incredible to attribute such ends to random search mechanisms, known by experience to be incapable, at least in finite time, of achieving even the simplest of ends [not even wooden spoons].' [194]

'Denton notes that Darwin himself had profound doubts, at times, over the enormity of his claims. Darwin wrote:

> 'Although the belief that an organ so perfect as the eye could have been formed by natural selection, is enough to stagger anyone . . . I have felt the difficulty far too keenly to be surprised at others hesitating to extend the principle of natural selection to so startling a length.'[195]

'In 1885 the Duke of Argyll recounted a conversation he had with an elderly Darwin,

> '[I]n the course of that conversation I said to Mr. Darwin, with reference to some of his own remarkable works on the *Fertilisation of Orchids,* and upon *The Earthworms,* and various other observations he made of the wonderful contrivances for certain purposes in nature—I said it was impossible to look at these without seeing that they were the effect and the expression of Mind. I shall never forget Mr. Darwin's answer. He looked at me very hard and said, *"Well, that often comes over me with overwhelming*

---

[194] Denton, M. 1985. Evolution: A Theory in Crisis. Burnett Books.

[195] Darwin, C. 1872. *On the Origin of Species,* 6th ed, 1962, New York: Collier Books, p. 41.

force; but at other times," and he shook his head vaguely, adding, "it seems to go away."' (Argyll 1885, 244).[196]

'Darwin was right to have his doubts, for, as Denton says, the only way he could have answered them would have been to provide rigorous probability estimates to show that the paths to such organs as the mammalian camera eye, the blood clotting cascade, the biochemistry of vision, bird's lungs, the metamorphosis of the caterpillar, and countless other biological mechanisms, could have been found by pure chance in the time available. Denton writes,

> 'Yet nowhere in the *Origin [of Species]* is any attempt made to provide quantitative support for the grand claim of the all sufficiency of chance [as a supreme creative principle]. It is true that Darwin appealed on many instances in the *Origin* to the enormous periods of time available to the evolutionary search but, as is the case with any other random search procedure, *time in itself tells us nothing of the probability of achieving any sort of goal unless the complexity of the search can be quantified.*[197]

'Interestingly, at the Wistar conference in 1966,

> 'Some mathematicians did try to make the calculations. The mathematician D.S. Ulam argued that it was highly improbable that the eye could have evolved by the accumulation of small [generally harmful genetic] mutations, because the number of [random] mutations would have to be so large and *the time available was not nearly long enough for them to appear.*
>
> 'Sir Peter Medawar and C.H. Waddington responded that Ulam was doing his science backwards; the fact was that the eye *had* evolved and therefore the mathematical difficulties must only be apparent. Ernst Mayer observed that Ulam's calculations were based on assumptions that might be unfounded, and concluded that "somehow or other by adjusting these figures we will come out all right. We are comforted by the fact that evolution *has* occurred." [198]

'That's intriguing' said Alisha, 'mathematicians pointed out, years ago, that Darwin's theories could not account for the eye but, because their calculations conflicted with his ideas, it was the mathematicians who 'had to be' wrong."

---

[196] Stanford Encyclopedia of Philosophy: Notes to Teleological Arguments for God's Existence, Note 44

[197] As W. Dembski does in *The Design Inference* (1998), *Intelligent Design Uncensored* (2010, with Witt), S. Meyer in *Signature in the Cell* (2009).

[198] Johnson, P. 2010. *Darwin on Trial.* IVP Books, p. 60, emphasis and text in square brackets added.

'Yes, the Darwinian's *beliefs* overrode the evidence. Darwin's entire theory is based on random causation as its supreme creative principle. Yet there is *no* evidence from any time or place that intelligent and purposeful mechanisms can arise from unintelligent and purposeless causes. Can you think of any?'

'You know, Ali, when it comes to thinking about origins, we are at an intellectual impasse today *not* because there is nowhere to go, but because those who currently control science are, in the main, materialists who insist that material nature is all there is or all that there can be.

'This means they *can* only resort to the regular laws of physics + lucky chemical chances, with no aims in mind, for these have no mind, to explain their own and everyone else's minds! This is why they must futilely argue, as they so futilely do, that nothing in nature, including their own minds and purposes, has any intelligence or purpose whatsoever.'

'Isn't that self-contradictory?' Mused Alisha. 'Should we take seriously the statements of those who say there is no purpose to anything in nature including, presumably, their own comments and conclusions about nature!'

'I agree, Ali. What they, also, do not allow for, is that it is really very easy to demonstrate that no complex living thing, let alone intelligent minds able to investigate nature, could ever arise by such mindless means.

'We know this! To keep appealing to endless chemical flukes of the gaps, because the alternative – intelligent or supernatural causation – threatens a rather weirdly cherished belief system, is a betrayal of science, the very purpose of which is to find out what is *true* and a key part of whose very method is supposed to be *to keep an open mind.*'

'Secondly, for materialism to continue to seem true, materialists must ridicule, dismiss or ignore every single spiritual, religious, mystical, psychic, supernatural, near death or other out of body experience anyone has ever had. Because, if they do not do this, their entire belief system will collapse.

'Similarly for the *intelligent causation* of living things. Nature, especially living nature, is *pervaded* by incredibly complex and purposeful functioning.

'Powerful evidences for intelligence and design or teleology abound: from the bat's wing to the whale's sonar, from the eagle's incredible eye to the human mind, from the cell's wall on the outside to the DNA on the inside which CODES for the very same wall which protects it from the outside.'

'Yet, right now, we are expected to fill all our knowledge gaps with brute stupid luck and to consider ourselves intellectually satisfied with this unreasonableness. Intelligence and purpose are not allowed because, to do so, would allow a spiritual foot in the door.'

We walked on.

19 Can Chance Create Software?

'The world's fossil records, Ali, show great changes over time. and we can call those changes evolution if we wish. Albeit, they do not match the kinds of changes which Darwin was predicting. However, when it comes to the intracellular level, there is no evidence for any kind of evolution at all.

'Molecular biology has also shown that the basic design of the cell system is essentially the same in *all* living systems, from bacteria to mammals. In *all* organisms the roles of DNA, mRNA and protein are identical. The meaning of the genetic code is also virtually identical in *all* cells. The size, structure and component design of the protein synthetic machinery is practically the same in *all* cells. In terms of their basic biochemical design, therefore *no living system can be thought of as being primitive or ancestral with respect to any other system, nor is there the slightest empirical hint of an evolutionary sequence* among all the incredibly diverse cells on earth. For those who hoped that molecular biology might bridge the gap between chemistry and biochemistry, the revelation was profoundly disappointing.'[199]

'So what do we do? Many still cleave to the view that there is no intelligence to the way living things come to be. The current mainstream scientific view is that, as Valtaoja said to Leisola, "Life is nothing else than physics and chemistry—mere electricity. There is," he says, "no reason to assume anything supernatural." [200] It is all 'unintelligently designed.'

'From the ribosomes to the proteasomes, from the amazing mammalian eye, to birds' feathers and birds' lungs, from our fascinating senses to our astonishing brains, from the bee's sting to the eagle's wing, it's all accidental and purposeless. This position is mainstream and, currently, the most influential voices in the science community incline to materialism, which means the perennial denial of anything spiritual or intelligently causal.

'Yet, despite these common views, some of the more agile intellectuals of our time are beginning to wonder whether Darwin's scheme really works?

'Gradually, they are rediscovering the classical understanding that, far from being mindless, living processes reflect amazing intelligence. For example, the

---

[199] Ibid.

[200] Lesiola, M. 2018. *Heretic: One Scientist's Journey from Darwin to Design,* Discovery Institute Press.

late Antony Flew, a famous atheist, who after decades of very public atheism, decided to 'follow the evidence,' as described in his book, *There is a God: How the World's Most Notorious Atheist Changed His Mind.*[201]

'What caused him to have a rethink?' Wondered Alisha.

'Among other things, the DNA CODED SOFTWARE in cells which, he realized, could *never* arise by chance. Codes, remember, are abstractions, they're not material things. Life's software, although named after, and written with, deoxyribonucleic acid, is abstract. It's pure information. DNA is merely the chemical ink by which some of life's *instructions* are conveyed.

'So it's as though computer languages like JAVA were called Silicon Chip Code or SCC?' Mused Ali.

'Yes, the term DNA CODE can be confusing. It might be more accurate, some have suggested, to let DNA stand for "Designed Natural Algorithms." [202] Because codes and instructions always consist in purposeful, non-random information. They cannot arise by 'purposeless chance.' Meaningful information which can be conveyed in many forms. Remember software is not *literally made* of the media by which it is transmitted. It is *immaterial and informational.* Even the world famous materialist, Richard Dawkins, acknowledges this. Here is this rather interesting quote by Dawkins:

'What has happened is that genetics has become a branch of information technology. It is pure information. IT'S DIGITAL INFORMATION. It's precisely the kind of information that can be translated digit for digit, byte for byte, into any other kind of information and then translated back again.' [203]

'Although Dawkins does not seem to notice, – perhaps because he is so averse to various traditional religious conceptions of the spiritual and the supernatural which he thinks are delusional, – this is precisely the kind of information which is *only ever known* to have and *can* only have intelligent causes. Remember, it's not meaningless information, which is sometimes referred to as *Shannon Information* and *can* be generated randomly.

'As Stephen Meyer explains in *Signature in the Cell: DNA and the Evidence for Intelligent Design,* an important book which explores the enigma of the origins of Life's Digital Codes, no amount of randomly tumbling coding symbols around will ever create INSTRUCTIONS OR SOFTWARE. Random

[201] Flew, A. 2009. *There is a God: How the World's Most Notorious Atheist Changed His Mind* Harper One.

[202] The Hottest New Computer Is: DNA www.evolutionnews.org/2017/10/03

[203] Richard Dawkins. 2008. Life: A Gene-Centric View, Craig Venter & Richard Dawkins: A Conversation in Munich. *The Edge*. Emphasis added.

processes can create meaningless, Shannon information but nothing functionally useful. In fact, so sophisticated are Life's Codes that there is, now, research in the use of DNA for ultra high density data storage and computing.

'*Futurism.com* announces, "Self-Replicating 'DNA Computers' Are Set to Change Everything." ... researchers at the University of Manchester are "working on turning strands of DNA into the next basis for computing." Because DNA's double helix can take two paths simultaneously, it can solve problems quicker.' [204]

'So why don't more people accept all this, like Flew, and move on?'

'Because it upsets the by now very widely established faith that science can explain everything in *purely materialistic* terms and that Life is not truly miraculous in origin. All things are just chemical coincidences.

'But why there is such an urge to de-intelligence and de-purpose nature in modern scientific thinking?' Ali looked sad. 'Why the desire to positively *suck the mind and the Soul, the magic and the enchantment out of everything?*'

'I agree, Ali, it is disappointing, and it's odd because you'd think scientists would be interested to discover that Life really cannot be pure accident. But although a classical spiritualist hypothesis of reality makes better sense of the data overall, including the modern – *non-religious* – evidence for supernatural phenomena of various kinds, it might lead to some loss of status for today's high priests of matter, physical scientists.

'You see, in the 19th century, in a dramatic intellectual reversal, it was physical scientists, like Darwin, who replaced the philosophers and theologians in the western world as arbiters of the most important truths and their intellectual descendants have little wish, now, to give up that glittering prize.

'Observe how many people today hang onto every pronouncement of famous science thinkers like Richard Dawkins and the late Stephen Hawking and how relatively few take much notice anymore of what someone like the Archbishop of Canterbury may have to say. It's a tremendous role reversal.

'Yet, take heart because, it's gradually becoming clear, even to some former materialists, like Antony Flew, that Darwin's hypothesis that Life might just be an accident which evolved by accident doesn't work.

'Thomas Nagel's book *Mind and Cosmos: Why the Materialist Neo-Darwinian Conception of Nature is Almost Certainly False* reflects this same gathering trend. He writes that a science 'heavily dependent on *speculative* Darwinian explanations of practically everything,'[205] opposed to the possibility

---

[204] See further at www.evolutionnews.org/2017/03/hottest-new-computer-dna

[205] Nagel, T. 2012. *Mind and Cosmos.* Oxford University Press; p.127.

of intelligent causation, as a matter of method and philosophy, is at an intellectual dead end. Such a science cannot provide a full explanation of reality, especially not of Life, Consciousness or Mind. He writes,

'in light of how little we really understand about the world' it would be an advance if the scientific establishment could liberate itself from 'the materialism and Darwinism of the gaps – to adapt one of its own pejorative tags.'[206]

'He argues that such a science is "incapable of providing an adequate account, either constitutive or historical, of our universe." He dismisses the materialist view of Nature which "purports to capture life and mind through its neo-Darwinian extension' pointing out that it is, 'antecedently unbelievable – a heroic triumph of ideological theory over commonsense . . . [adding that he] would be willing to bet that the present right-thinking consensus will come to seem laughable in a generation or two . . ." [207]

'Leading modern intelligent design thinker Stephen Meyer writes,

'During the last forty years, molecular biology has revealed a complexity and intricacy of design that exceeds anything that was imaginable during the late-nineteenth century [the steam-engine age to which Darwin belonged].

'We now know that *organisms display any number of distinctive features of intelligently engineered high-tech systems:* information storage and transfer capability; functioning codes; sorting and delivery systems; regulatory and feed-back loops; signal transduction circuitry; and everywhere, complex, mutually-interdependent networks of parts. ... the materialistic [steam-age] science we have inherited from the late-nineteenth century, with its exclusive conceptual reliance on matter and energy [cannot] account for the biology of the information age.' [208]

'Nowhere, in the real world, Ali, do micro-miniaturized manufacturing systems, run by very sophisticated software, self-assemble by 'luck.' This is why Stephen Meyer points out that scientists are called on to adopt the most *reasonable* explanations for a given phenomenon, not to avoid them. This is why, when confronted with living things which display many "features of intelligently engineered high-tech systems," to rule out intelligent causation from the start is not only unreasonable, it's unscientific.

---

[206]   Ibid.

[207]   Ibid.

[208]   The Origin of Life and the Death of Materialism by Stephen C. Meyer, Phd. The *Intercollegiate Review* 31, no. 2 (spring 1996). Text in square brackets added.

'As Meyer explains, students of natural history, like Darwin, who cannot do lab experiments to recreate the past, need to choose between various competing hypotheses in their attempts to make sense of life. The hypotheses they choose ought to be the ones with the best explanatory power and the ones most inline with their real-world knowledge of causes and effects.

'Purely physical causes, like wind and rain, produce *randomly* complex effects, like landscapes. They are, though, causally inadequate to create highly ordered complex mechanisms, like living things. Only intelligence can create the kind of purposeful, *non-random* complexity characteristic of living things, and everything we make. A complexity always dependent on intelligently designed parts assembled in specific (non-random) sequences.

'We know, by now, that there is nothing in the laws of unguided chemistry to produce *living* things. This is why the 'aim of the modern movement in biology . . . to explain all biology in terms of physics and chemistry,' [209] favored so a priori by thinkers like Monod and Crick, doesn't work.

'This is why Thomas Nagel writes that the materialist view of Nature is, 'unbelievable – a heroic triumph of ideological theory over commonsense.' [210]

'This is why, as Stephen Meyer explains, "Considerations of causal adequacy" provide us with real-world, experience-based criteria "by which to test, accept, reject, or prefer competing historical theories" [211] which attempt to explain unwitnessed past events.

'This is why, when a theory, like intelligent design, cites causes, like intelligence, which are known to produce the effect in question, they meet the test of causal adequacy. Darwin's laws of *unintelligent* design – the laws of mindless chance – fail the test. They demonstrably cannot create codes or software. They are demonstrably inadequate to create Life or Life's Codes.

'Stephen Meyer writes,

'because experience shows that an intelligent agent [or agents] is [are]. . . ***the only known cause of specified, digitally encoded information***, the theory of intelligent design developed in this book has passed two critical texts: the tests of [1] causal adequacy and [2] causal existence . . .' This is why intelligent design 'stands as the best explanation of the DNA enigma.'[212]

[209] Crick F. 2004. *Of Molecules and Men.* Prometheus, p. 10.

[210] Ibid.

[211] Meyer, S. 2009. *Signature in the Cell.* Harper One; p. 405, emphasis added.

[212] Ibid. p. 405, emphasis and text in square brackets added.

'Remember, regular physical laws, by themselves, cannot create the SPECIFIC, PURPOSEFUL SEQUENCES OF INSTRUCTIONS found in all codes and software. No amount of randomly tumbling various coding symbols around will ever generate anything meaningful.

'Michael Polanyi, the distinguished chemist, pointed out that if the precise sequences of the letters in Life's Codes were *physically* determined, by the regular bonding laws of chemistry, "then such a DNA molecule would have no [meaningful] information content ... [The sequence of the DNA letters] must be as *physically indeterminate* as the sequence of words is on a printed page." [213]

'This is why a regular law which can generate simple, regular sequences along the lines of ABC+ABC+ABC, which can describe some parts of nature, like crystals, cannot generate the *specific, non-regular, sequences* required by instructions like the DNA CODED sentence, 'TTAAGGCC ATAGGAGC CAACGGTT,' or one of our sentences like, 'P L E A S E   F E E D   T H E C A T.'

'Meyer explains that the only agencies which can create software and instructions are *mind and intelligence*. So, although the ultimate sources of nature's designs cannot be seen physically, they can, reasonably, be inferred.

'He makes a powerful case for intelligent design. It is the failure of our current science *not* to recognize the purpose and intelligence in Living things which is so peculiar, not the ideas of those like Meyer. Thinkers able to see the strict limits of purely materialistic approaches in scientific inquiry.

'But don't some say that intelligent design is a dead end? Because ID is like arguing, "God does it all, so let's just give up on our inquiries and go home."

'Yes, Ali, they do, but they are mistaken. ID is a hypothesis in science, just as Darwin's theory of unintelligent design or UD is. Both hypotheses are scientific because they can be *tested*. Our word science comes from *to know* in Latin. It does not come from the Latin for: "To think materialistically."

'Conceding that there may, after all, be *intelligence and purpose* behind living things doesn't end science nor does it endorse any particular set of religious or spiritual views. To assume that, "If intelligence is involved, we'll never be able to work out how living things come to be," is defeatist. If the Life Forms, including we, have intelligent causes, we may, gradually, work out more about them.

'If we say a river valley or other landscape is unintelligently designed we are not being controversial because we can *observe* how it has formed. While there may be intelligent and purposeful laws behind the river and its valley, behind the hydrological cycle driving it, and behind our Star, the Sun, driving

---

[213] Michael Polanyi, "Life's Irreducible Structure," *Science* 160 (1968), 1308–12, quoted by Wells, J. 2006. *Darwinism and Intelligent Design*. Regnery Publishing; pp. 100–101, emphasis added.

that, its immediate forms are random, so we can, reasonably, call it 'unintelligently designed' or UD.

'But, when it comes to *living* things, to argue that unguided physical forces can create them is impossible to justify. Perhaps both this lovely valley, and the sheep gently grazing these hillsides, derive, ultimately, from a supernatural source – an amazingly powerful and creative *final Source* to which humanity gives its traditional religious names.

'But, while we can, reasonably, describe this landscape as *unintelligently* designed, can we usefully apply the same reasoning to the existence of the sheep or of we who are here to observe it and to think about it?

'If we set a water pipe running on a steeply arranged pile of earth it would soon create a miniature landscape of rivers and valleys below. But we would *not* have consciously designed them. They would form randomly. But if we take rocks and oils out of the ground and, after thousands of hours of intelligent effort, turn them into something genuinely clever, like a Computer, Car or TV, to call them 'unintelligently designed,' that is due solely to pure chance + regular physical laws, would be intellectually unreasonable, not to say absurd. The same is true for *living* things, only many times over.

'Science is about finding out what is *true*. It's not about clinging to an a priori ontology, to a predetermined worldview. In the 1500s, at the dawn of the modern age, Sir Francis Bacon urged scientists to look to *physical* causes, as much as possible, when seeking to explain something in nature, especially at the start of their investigations. The approach which we now know as *methodological* materialism or methodological 'naturalism.'

'But Bacon, unlike Darwin and his modern super fan, Richard Dawkins, was not a *philosophical* materialist. He did not insist that the *only* possible explanation *must* be a materialistic one. He did not dismiss the multi-dimensional, the supernatural and the spiritual out of hand, as so many do today, in the name of 'rational science.' He would have found such a position deeply *irrational!*

'If materialist thinking cannot give a *full* account of reality, if there is *more* to us than matter, and if there are good, modern ways to demonstrate this, our sciences should allow for it. Bacon would not have disagreed with this.

'These are demanding issues, it's true, but should we not be interested in them? What, after all, is more important than knowing *Who and What* we really are? Are we just cosmic accidents or something much more interesting?'

We walked on in the lovely valley.

20 The Trouble With Hume

'Contrary to the claims of its detractors, Ali, the central concept of modern intelligent design thinking is not, specifically, that living things cannot evolve. It is, rather, that they can *neither arise nor evolve* by purely mindless means.

'No one knows whether an intelligent Higher Nature, which 'just is,' to which humanity gives its traditional religious names, can *create* but cannot *evolve* things. As this cannot, currently, be tested, it's a matter for theological debate, not for science.

'It is not, though, solely a theological matter to wonder whether there may be *more* to reality than the purely physical. Even Richard Dawkins argues that the Supernatural, up to and including God or gods, can be a hypothesis of science. He just feels that it's one which has been tested and disproved.

'Although they disagree with Dawkins's materialism, none of today's well known advocates of intelligent design, like Behe, Meyer and Dembski, are claiming that the modern evidence for ID is proof for their particular religious views. No, their only claim is that, there is, demonstrably, far more going on with Living things than can be explained by the wholly materialistic approaches which thinkers like Darwin, and modern science as a whole, insist are all we can rely on.

'If our sciences became more open once again to wondering whether visible, physical nature, and the natural laws we know of so far, may be derived from an intelligent underlying or overarching *spiritual nature,* it would be uncontroversial to say that all things have *'natural'* explanations because such a hypothesis would allow for the more subtle elements of reality.

'But if we decide, a priori, that nature is physical only and there's *no* intelligence to *any* of it, including even to our own complex bodies and clever minds, we are hardly going to wonder when and in what circumstances scientists may be looking at intelligent and purposeful processes, as can be seen in living things, versus mere chance, as in the making of landscapes, where nothing beyond nature's regular laws + raw chance need be involved.

'But, for those of us who do not conflate science with materialism, who do not perceive science to be in a win-lose *'struggle with the supernatural,'* for those of us unwilling to leave our brains at the door and to *"accept scientific claims that are against common sense,"* for those of us wary of *"unsubstantiated just-so stories,"* and who do not have *"a prior commitment to materialism,"* as Lewontin put it, it is not unreasonable to wonder whether intelligent causation may be part of reality, to argue for it and to show it.'

'So how do you do that?' Asked Alisha.

'One way is to use philosophy and logic. For example, is it *logical* to argue that *mind* is an accident of *mindless* matter? Isn't this self-refuting? Is it *rational* to hold that human beings are just soulless bio-robots, which, unlike all robots in the real world, 'accidents' built? Just randomly swirl the parts about for long enough and all will be well? Let's put all our *real-world* experience of robot-building on one side? Is this a scientific approach?

'Secondly, quantitative and statistical arguments: not only are codes and software never known to arise by chance, they *cannot* do so. Yes, we can still ask, as a somewhat pointless thought experiment, "Could Life's complex DNA CODES, unlike all other CODES, arise by chance?" But should we take no notice of the answer? *"No, it's impossible, the laws of probability forbid it."*

'Thirdly, empirical arguments. We know accidents of physics can produce randomly evolving river valleys and forest fires. They are *never* known to create anything remotely intelligent or mechanically functional. To suggest otherwise, as an 'unsubstantiated just so story,' because 'we have a prior commitment to materialism,' makes our science of origins look unreasonable.

'Prior to Darwin a *wholly* materialistic way of looking at nature was unusual and intelligent causation was taken for granted. For example, a couple of generations before Darwin, the philosopher and theologian, William Paley, pointed out that anyone who came across a *clock* lying on the ground, as compared to a random rock, could easily work out it was designed and not some random thing. He argued, by analogy, that the Life Forms were mechanisms with machinelike characteristics which could not be mere 'natural accidents' of chemicals anymore than a watch could be a very lucky but entirely 'natural accident' of its components. He wrote,

> 'Every indication of contrivance, every manifestation of design, which existed in the watch, exists in the works of nature; with the difference, on the side of nature, of being greater or more, and that in a degree which exceeds all computation.' [214]

'No one would question Darwin's paradigm of pure unintelligent design if the way we made watches, cars and computers – assembled, like all living things, in *very specific and pre-determined sequences* of parts – was to put crumbled rocks and oils in containers, set them on a tumble, then, later, take out the finished products. Yet, we now realize, all the Life forms are far, far

[214] Paley, W. 1802. *Natural Theology*

more complex than *anything* we make, using all our intelligence, so to imagine they might arise by the mere swirling-about-of-chemicals-for-a-long-time + regular physical laws + huge-amounts-of-irregular-luck is no longer tenable.'

'So what happened to Paley's design arguments?'

'They were, in time, dismissed by the mainstream science community partly because of its over credulous enthusiasm for Darwin and partly because of the philosopher David Hume's claim that living things are not strictly comparable to machines and, therefore, he felt, Paley's arguments were invalid. While Hume seems not to have noticed that the whole *point* of arguments from analogy is that they are not intended to compare precisely like with like – if they were, they'd be pointless – it is true they can only go so far. In this case, however, their limitations *hinder* rather than help Hume's own argument.'

'Why?' Asked Alisha.

'Because living things – while they can be compared to machines in some respects – are *far, far more complex* than any machines we make. Yet no machine ever arises by chance. So how could nature's machines do so?

'Secondly, because Life's digital instructions are not *'like'* CODE which would be an argument from analogy, and subject to the limitations of such arguments, they *are* CODE. Code far more sophisticated than any we make.

'This is why Bill Gates writes of Life's DNA software that it is,

> 'like a computer program but far, far more advanced than any software .
> [we've] ever created,'[215]

'This is one of the disappointing things with materialist thinking in science, Ali. Not only can it be self-contradictory but, at times, it lacks integrity...'

'Go on...'

'Well, one minute it tells us that Nature is *purely mechanical* and her many complex mechanisms are just coincidences, without purpose or meaning. But if you point out that machinery, simple or complex, or the software to drive it, *never* arises by chance, that this is *a contradiction in terms,* your design argument suddenly becomes invalid because, according to Hume and his followers, you are arguing from analogy and the Life Forms are *not,* after all, anything like machinery. This, though, is not merely tricky, it's false.

'Because, ironically, just as so many rushed to adopt Darwin's proposals that raw chance could explain all, the scientists of his day had no idea just how machine-like some biological mechanisms would, ultimately, turn out to be.'

'So William Paley was eventually vindicated?'

'Yes, and to an astonishing degree. Eastman and Missler write:

---

[215] Gates, B. 1996. *The Road Ahead.* Penguin. p228.

'However, the astonishing discoveries in molecular biology during the last 40 years have finally and unequivocally demonstrated that living systems are, in fact, machines – even to the deepest, molecular level! From the tiniest enzyme to the most complex organ systems found in man, Paley's machine analogy is confirmed.

At the enzymatic level we see an eerie resemblance to the design and operation of chemical factories. At the organ level we find "hardware" of an unimaginable complexity and ingenuity. In our five senses we find sensory receivers made of multiple components, each machine-like, the operation of which is absolutely necessary for each sense (taste, sight, smell, hearing, touch) to function properly. In the function of the human heart we see an incredibly efficient and durable hydraulic pump, the likes of which no engineer has imagined. Finally, in the structure of the human brain we find a computer 1000 times faster than a Cray supercomputer with more connections than all the computers, phone systems and electronic appliances on planet earth! In each of these systems, at every level, we find machine-like structures which are truly "teleonomic" (purposeful) aggregates of matter, each executing its role in a pre-programmed manner.' [216]

'In *Evolution: A Theory in Crisis,* molecular biologist Michael Denton makes a similar point. He writes,

'It has only been over the past twenty years with the molecular biological revolution and with the advance in cybernetic and computer technology that Hume's criticism has been finally *invalidated* and the analogy between organisms and machines has at last become convincing . . . In every direction the biochemist gazes, as he journeys through this weird molecular labyrinth, he sees devices and appliances reminiscent of our own twentieth-century world of advanced technology. *We have seen a world as artificial as our own and as familiar as if we had held up a mirror to own machines* . . . Paley was not only right in asserting the existence of an analogy between life and machines, but was also remarkably prophetic in guessing that the *technological ingenuity realized in living systems is vastly in excess of anything yet accomplished by man.*'[217]

'You know, Alisha, the difficulty modern ID theory poses for Darwinian thinking is not a lack of evidence for intelligence or design in Nature – for

---

[216] Eastman, M C. Missler. 1995. *The Creator Beyond Time and Space.*

[217] Denton, M. 1985. *Evolution: A Theory in Crisis.* Burnett Books; p. 341, emphasis added.

these are pervasive. Nor that it poses any real threat to practical, everyday science. No. The problem is *philosophical and ideological.*

'ID challenges *materialism and so called naturalism.* Those who control science today are very opposed to the classical idea that the *super* natural, that some underlying or overarching intelligence or intelligences could play any part in the arisings of nature. And, even if it does, they oppose the idea that contemporary scientists will be capable of detecting this or of elucidating it in any way.

> "In the cerebral cortex alone, there are roughly 125 trillion synapses, which is about how many stars fill 1,500 Milky Way galaxies. ... Researchers at the Stanford University School of Medicine ... [have] found that the brain's complexity is beyond anything they'd imagined, almost to the point of being beyond belief, says Stephen Smith, a professor of molecular and cellular physiology . . . One synapse, by itself, is more like a microprocessor ... In fact, one synapse may contain on the order of 1,000 molecular-scale switches. A single human brain has more switches than all the computers and routers and Internet connections on Earth." [218]

'This, Ali, is not good evidence for unintelligent design! Anyone who disagrees is not thinking. No micro-processor we make ever arises by pure chance. Yet our present science asks us to believe precisely this. This is why, currently, the brain's amazing organization is held to arise, in essence, from chemical coincidence (even though this is impossible), to be good evidence for unintelligent design (which is unreasonable), for meaningless functionality (which is an oxymoron) and for pointless purposefulness (which is another).

'Unfortunately, this illogical Darwinian ideology of mindless, random causation has made important parts of our sciences hostages to nonsense and into vehicles, not for enlightenment, but for academic dissembling and downright unreasonableness. For, if we may not call such processes intelligent and purposeful, our very language has become meaningless and key parts of our science ways to distort our understanding of reality, not explain it to us.

'Why do you think this has happened?'

'Well, it's partly because of the confusing conflation of various ideas and discoveries, some valid, others false, leading, eventually, to the intellectual mess we find ourselves in today, where, *irrationally,* mindless chance has been elevated to a supreme creative principle when it comes to the construction and functioning of living things.'

'What are those ideas or discoveries?' Asked Alisha looking intrigued.

---

[218] Moore, E. A. 11.17.2010. cnet.

1. Conflating the *scientific method* of empirical hypothesis testing with the materialist *belief system*. This means that if an explanation for something, like the Cell or Heart or Brain, isn't *wholly* materialistic it is considered unscientific. This is why, currently, all ideas of *intelligent design* are dismissed. Yet, people forget, it is not scientifically established that the supernatural is unreal and that all living things are mere chemical coincidences.

'These are the very questions in need of answers: Could living things arise by mere chance? Is materialism true? Can it be shown to be? Is science dependent on materialism for its success? Bruce Greyson, a leading expert on near death and out of body experiences, writes,

> 'Materialists often claim credit for the scientific advances of the past few centuries. *But it is the scientific method of empirical hypothesis testing,* rather than a materialistic philosophy, that has been responsible for the success of science in explaining the world. If it comes to a choice between empirical method and the materialistic worldview, the true scientist will choose the former'[219]

2. The discovery that Life on earth is far more ancient than Bishop Ussher's 17th century reading of the bible led him to believe, followed by the *non sequitur* that there can be no intelligent causation or meaningful creation at all.

3. Not realizing (a) that an accurate description of natural history – including the fossil and other evidence – is not the same as (b) Darwin's theory which tries to explain this evidence, nor (c) the same as concluding that only a *materialist* explanation, like Darwin's, can make sense of it versus (d) the classical proposal that an explanation which allows for supernatural causation, or for intelligent design, may be more true and more complete.

'This muddled thinking means trying to make sense of a confusing mixture of propositions, some of which are true and some false:

1. The biblical creation story in Genesis is *not* science. *True,* it's metaphor.
2. So something like Darwin's theory is the only alternative. *False.*
3. Life on Earth is extremely ancient, the fossils prove it. *True.*
4. So a *super* natural creation or intelligent evolution is not possible. *False.*
5. Life on Earth has changed greatly over great time. *True.*
6. This proves Darwinian macro-evolution is a fact. *False*: (a) The nodes on Darwin's tree of life, where his hypothetical common ancestors should be, are empty. Fossils for them ***do not exist***. (b) The vast numbers of transitional

---

[219] Greyson, B.: "Commentary on 'Psychophysiological and Cultural correlates Undermining a Survivalist Interpretation of Near-Death Experiences.'" *Journal of Near Death Studies* 26, no. 2 (Winter 2007); P.142.

species called for by his theory *do not exist – neither fossil, nor living. (c)* There are continuities of design, as with our designs, but *no evidence* for prior species slowly turning into new species *either by random, or by intelligent means.* (d) Not just the fossil evidence, but that from embryology, genetics and biochemistry all falsifies Darwin's theory.

4. Darwin's theory or something like it has to be true because *materialism* is true, science has proved it. *False.* Materialism is, like Darwin's theory, a hypothesis which may or may not be true. The data shows it to be *false.*

'Yet, despite all this, or, maybe, because of it, it takes a lot of courage to speak up for any ideas of intelligent design in mainstream science today.'

'Why?'

'Because the bias towards Darwinism and materialism is so entrenched that those who question these are accused of being unscientific and this puts their reputations and careers at risk.'[220]

'Look up intelligent design in wikipedia, for example, and you'll be falsely told that it has been successfully rebutted by the science community.

'This is not true. It *is* true that ID proposes alternative ideas to Darwin's and it is also true that it is, currently, a minority view. But ID has not been falsified and majority views do not, by themselves, amount to scientific truth. Science is not just about majorities and consensus but also about competing ideas.

'At the least, ID is an interesting – and testable – hypothesis of science, just as Darwin's theory of unintelligent design or UD is. Science is about finding out what is true. So, was Darwin right? Is materialism true? Could Life be a mere accident? Is this possible? Is there any evidence for it? Or, is there *demonstrably* more to Life and Mind than mere chance? This is the hypothesis behind modern ID thinking and, until Darwin's time, it was the mainstream scientific view. But, today, the wikipedia entry on ID starts,

> 'Intelligent design (ID) is the *pseudoscientific* view that 'certain features of the universe and of living things are best explained by an intelligent cause, not an undirected [and purposeless] process such as natural selection.''

'So, according to wikipedia it's 'pseudoscientific' to even *wonder* whether intelligence could play any part in the creation of reality and whether scientists might be capable of noticing this? Does this mean, then, that all the great philosophical and scientific minds – from Plato and Aristotle to Galileo and Newton, from Leibniz and Boyle to Faraday, Maxwell and Planck – who *all* believed in intelligent causation in some form or other were pseudo scientists?'

'It does seem to be a very narrow position.' Mused Alisha.

---

[220] Wells, J. 2006. *The Politically Incorrect Guide to Darwinism and Intelligent Design.* Regnery Publishing.

'Yes, Ali, it is. And it is misleading because it is *materialist prejudice masquerading as science*. But wikipedia does get one thing right. As it says, natural selection is a mindless and 'undirected process.' Very true. But for getting Life started *or* evolved it is akin to putting chemicals in a barrel, randomly swirling them around, then hoping something clever will emerge, over and over again. This, according to wikipedia, is the 'scientific' position.

'This is why random causation which – with no purpose, with no search criteria pre-defined, with no aim or goal in mind – is *not capable, in the real world*, of assembling even the simplest of functionally working things, not even wooden toys, is held up by wikipedia as a supremely creative principle.

'Yet, Darwin could not find *one* real world example, *not one!*, to demonstrate the allegedly so very amazing *macro-evolutionary* powers of this principle. And no one has found any since.

'Yet, we are asked by wikipedia, and many otherwise reliable institutions, to treat modern ID thinking as *pseudoscientific* but the various theories of unintelligent design or UD, including Darwin's version, as 'proper science.'

'So it's scientific to hypothesize that entirely mindless causes can give rise to everything, even though this is demonstrably impossible? Even though even some very well known materialists, like Wald, Crick and Monod, admit this?

'But it is not scientific to wonder whether intelligent causation, no matter it be subtle or mysterious, could play any part in the creation of reality? In fact, to even think of doing so is a 'religious argument.'

'So, is it an 'atheist-materialist' argument to argue for unintelligent design? That pure, stupid, purposeless chance is incredibly creative? If so, it should not be taught in science classes all over the world, that's for sure.

'Most of humanity, including most of the greatest thinkers who ever lived, believes there is more to life than the purely physical yet, according to wikipedia (at time of writing), it is pseudoscientific for scientists to even wonder whether Life could possibly *arise* or *evolve* as a result of the hypothetically *wholly mindless and utterly unintelligent* processes into which materialist thought puts so much *faith?* Wikipedia continues,

> 'Educators, philosophers, and the scientific community have demonstrated that ID is a religious argument, a form of creationism which lacks empirical support and offers no tenable hypotheses.'

'This, too, is misleading. Just because intelligent design posits that there may be more to Life than pure chance does not make it a 'religious' argument.

'Secondly, it is not true that modern ID thinking 'lacks empirical support,' nor is it true that it 'offers no tenable hypotheses.' Rather, these are all assertions of those who just don't like the idea. One could just as well say,

'Many philosophers, educators and scientists have demonstrated that Darwin's theory of unintelligent design (UD) is a modern atheist-materialist argument [it is often used in this way]. It is a modern creation myth for materialism [quite true] which lacks empirical support [very true] and it offers no testable hypotheses [false].'

'Whilst it's true that *Darwinian unintelligent design theory* operates as a mindless-chance-based creation story for the modern atheist–materialist belief system favored by thinkers like Crick, Dawkins and Lewontin, it *can* be tested, so it's not beyond scientific investigation. It has, though, been tested, again and again, and it has been found wanting, again and again.

'Darwin's theory is not borne out by the data and the idea that such an evolution, even if there *was* good evidence for it, could be driven by random genetic mutations, or by any other random process, is demonstrably untenable.

'From genetics to molecular biology, from fossils to embryology, Darwin stands falsified but, like a foolish emperor, we are asked to agree that he looks magnificent, to consider ourselves *intellectually satisfied* and to be grateful that he showed us that our existences are, basically, meaningless!

'It is also not true to say that modern ID theory cannot come up with any testable or worthwhile scientific hypotheses of its own. Here are just a few:

1. The ID hypothesis that Life could *never* arise by chance, no matter how much time or how many chances you allow, can be tested. The results are in and the answer is an emphatic: "No, Life by chance is impossible." The reluctance to accept the results is down to emotion and ideology, not science.

2. The ID hypothesis that the 'Junk-DNA' in Cells may not be 'junk' after all is interesting, it can be tested, it is being tested, and the results are in line with the predictions of modern ID thinking.[221]

3. The ID hypothesis that no software or coding system could ever arise 'by accident' can be tested. The results are in. The random causation of CODES and software is *impossible*. But the materialist claim that there are only unintelligent and accidental causes in or of Nature prevents their acceptance.

4. The Darwinian hypotheses that the incredibly complex biochemistries of cells, of vision and of the blood clotting cascade, among many others, could

---

[221] 'In conflict with the 'junk-DNA' hypothesis, the ENCODE project has determined that "The vast majority (80.4%) of the human genome participates in at least one biochemical RNA- and/or chromatin-associated event in at least one cell type." Tom Gingeras, a senior scientist with ENCODE, states that "[a]lmost every nucleotide is associated with a function." Junk No More: ENCODE Project Nature Paper Finds "Biochemical Functions for 80% of the Genome," Casey Luskin, September 5, 2012. evolutionnews.org.

arise due to the mindless, natural selection of chemicals can be tested, as Michael Behe and others have done, and it can be shown to be wanting.

'Yet, despite this, wikipedia's ID attacking article continues,

'Both irreducible complexity and specified complexity [theories in modern intelligent design] present detailed negative assertions that certain features (biological and informational, respectively) are too complex to be the result of natural [i.e. purely physical or random] processes. Proponents [like Behe, Myer and Dembski] then conclude by analogy that these features are evidence of design. Detailed scientific examination has rebutted the claims that evolutionary explanations [i.e. mindless chance + regular physical laws] are inadequate [to explain all natural phenomena].'

'More misleading statements from wikipedia! There is *no evidence* that Life's complex CODES could write themselves (Meyer, Lennox, Gitt). There is *no evidence* that Living Cells could *ever* arise by mere chance (Dembski, Hoyle, Eastman, Missler, Wald, Denton). Even some famous materialists admit it's impossible.

'There's *no evidence* that the incredibly complex biochemistry of vision could ever arise by mere chemical chance (Behe).

'There's *no evidence* that dinosaur lungs could randomly turn into avian lungs without fatal consequences for the creatures concerned, 'death within minutes,' (Denton).

'There's *no evidence* that the bombardier beetle's astonishingly sophisticated weaponry could arise by chance (Simmons). There's *no evidence* that the fabulous metamorphosis of the crawling caterpillar into the beautiful butterfly could happen by mere genetic coincidence (Wallace). There are never ending *assertions* that such things are possible but never any *evidence.*'

'So why do you think this keeps happening?'

'Because Life's deeper causes cannot be seen physically. They can only be known by intuitive means or inferred, using logical, deductive reasoning. And, because, in our often troubled and confused times, the materialist beliefs of famous thinkers and influencers like Darwin and Haeckel, Engels and Nietzsche, Marx and Lenin, Wald and Monod, Crick and Lewontin, Hawking and Dawkins, have become more fashionable than the non-materialist thinking of famous idealists like Plato and Galileo, Leibniz and Boyle, Newton and Planck.

'Science, though, is supposed to remain open-minded in its inquiries. It is not supposed to be a dogmatic belief system which cannot be questioned.'

We walked on.

21 Why Don't They Understand?

'One thing I find curious about the debate on origins, Ali, is why some people seem to find it so hard to understand intelligent design. Isn't ID just as interesting a hypothesis of Life as Darwin's notions of unintelligent design? Unless we already know Life to be a mere accident? The thing is, we don't.

'What we *do* know is that, impossible as it might seem existence exists. After all, *why* should there be anything at all, rather than nothing at all? No one can say. But the seemingly impossible has happened. Out of *"whatever-it-is"* comes everything. But is it an intelligent *"whatever-it-is"* or mindless?

'These are the two basic hypotheses: (1) Life has intelligent origins. (2) No, it's a chemical fluke. But, if you point out that hypothesis (2) is demonstrably impossible, the advocates of this view, like my old friend, Vicki, resort to comments like, "Well, *wikipedia* says intelligent design is pseudo-science."'

'So how do you answer that?'

'One answer is, "Have you actually studied intelligent design? Do you even know what it is?' To which she'd reply, "But it says in *wikipedia* that ID can't be a proper theory unless we know who the intelligent designers are." I'd respond, "Look, we don't need to know who designed a coding system or clock, or why, to work out that the code or clock could not, possibly, have arisen by mere luck."

'To which she'd reply, "We may not know, right now, how Life's CODES or Genes or Cells or Hearts or Brains, might arise by pure purposeless chance, by what you call 'unintelligent design,' but maybe it's possible?"

'"But no CODE can write itself! *We know this! We can demonstrate it, mathematically and scientifically,* using the very same quantitative and scientific methods which you claim to believe in. In any case, it has been shown again and again that Life could *never* arise by pure chemical luck. It is scientifically, chemically, physically, mathematically impossible."

'And the upshot was...?'

'Mostly just bafflement on my part. My old friend seemed unable to understand that there could be more than one hypothesis of Life. For her, it seemed, a 'scientific' explanation had to be purely *materialistic and random. Intelligence could not be considered.* Yet, when pressed on the demonstrably limited powers of random causation, she would say, "Well, it's not necessarily 'accidental,' it's just that there's so much we don't yet know."

'To which I'd reply, "Look, what we *do* know is that there are only two or three kinds of causes, (1) intelligent, (2) random or, (3), a mix of the two. And, for most of history, the hypothesis of intelligent causation was mainstream."

'"Obviously, if we could physically see *the subtle, intelligent forces* behind Life's CODES, the mammalian eye, the brain, the bombardier beetle's weaponry or the astonishing intracellular protein machines, not to mention our own sometimes quite clever minds, there'd be no debate. But we do not have to physically see the ultimate causes of an object or mechanism, be it a coding system, a heart or a brain, to be able to work out whether or not it is more likely to have an intelligent cause than a wholly mindless one."

'For example, in *Darwin Devolves,* Michael Behe, highlights the fascinating *gear-wheels* on the grasshopper's legs which enable it to leap so amazingly. *These cogs literally look completely machinelike.* I showed these to Victoria.'

'What did she say?'

'Her response was, "I've looked up Behe in *Wikipedia* and it says he's a pseudo-scientist who pedals religious views as 'science.'" Exasperated, I said. "You're a biochemist and you have worked in science. So why don't you read his book for *yourself,* see what he says, and give his ideas a chance.

"He shows, for example, that, contrary to the breezy claims made for their alleged creative capabilities, random genetic mutations do not have the power *either to create or to evolve* living things. They are genetic copying errors. They *degrade* living things. They are not, fundamentally, constructive."

"If, having done this, you remain unconvinced, at least you'll have given his ideas a fair shot, rather than relying on *wikipedia* and other often more or less biased reviewers who refuse to give Behe's arguments a fair hearing

"Science, is about trying to find out what is *true.* What is true is either that (1) Life and Mind have *mindless causes* or (2) *intelligent causes.* Both the ID theories of those like Behe and Darwin's UD theory are hypotheses about the *true* nature of reality, the very things which it is science's job to inquire into.

"Now, if it can be shown, as it has been, that so many key parts of Darwin's notions fail, we can return, reasonably, to William Paley's classical hypothesis that, if aimless luck could never build a Living Cell or the DNA Coded Software to help run it, let alone a Heart or Lung or Brain, no matter how much time we allow, maybe intelligent or supernatural forces can do so."

'Ah, Ollie, you care so much about all this.' Sighed Alisha.

'Yes. Because what could be more important than knowing whether we are mere accidents or something more? As it is clear by now that many key features of living nature *could not possibly arise by pure luck,* no matter how much time we allow, the logical alternative hypothesis is the classical one of supernatural causation – what, nowadays, some call intelligent design.'

'One thing, though, Ollie. While Darwin argued for *unintelligent design,* he did put forward a possible mechanism to account for it – random variations to living things, which, if their bearers survived, might lead to improved varieties or entirely new species. But today's advocates of ID don't have any proposed mechanisms as to how ID might occur, do they?'

'Hold on, Ali. Darwin had *no* real proposals as to how living things might arise, *by pure chemical luck!* Remember what he wrote to his friend Hooker.

'But if (and oh what a big if) we could conceive in some warm little pond with all sorts of ammonia and phosphoric salts – light, heat, electricity etc present, that a protein compound was chemically formed, ready to undergo still more complex changes'[222] [and, Darwin imagined, accidentally become a Living Cell, 'the most complex system known to man.'[223]]

'These kinds of speculations are no longer enough. Unlike Darwin, we now know that living things, even of the most basic kind, could *never* arise in this way. It's demonstrably impossible. Even some materialists admit this. They just do not like the scientific, philosophical and political implications.

'Which are ?'

'That there really is more to reality than the purely material. That the spiritually minded may not be mistaken in believing in deeper dimensions to reality. That our physical lives may not, after all, be all there is to us. Reports of Life after death may be real and so on. So there is a lot at stake here.

'As to Darwin's theory of macro-evolution, almost *none* of his key predictions are borne out by the data: the fossils, embryology, genetics or molecular biology. If all Darwin had predicted were micro-evolution or just great changes over great time, almost everyone would agree with him, because both are true. But he was trying to do much more than that.

'He argued that those changes were *non-supernatural, non-miraculous.* Now, if macro-evolution, as he predicted it, *really were true,* one could argue, as some Darwinian theists do, that Darwin had discovered God's way of creating. But that the evolution which he attributed to *random causes, to unguided, unintelligent design,* could be better explained by meaningful causes.

'You mean macro-evolution *did* take place but it was intelligently guided?'

'Yes. This is the idea adopted by some religious believers, old-earth creationists for example. The major problem with this is that there is *no*

[222] Darwin to J.D. Hooker. 2.2.1871.

[223] Davies, P. 1998. *The Origin of Life.* Penguin.

evidence to support macro-evolution, as Darwin described it, his core idea of one species changing into another and then another, over and over again.

'So why continue, Ali, to try to account for a process, whether we conceive it to be random, as Darwin argued, or intelligently directed, as god-believing followers of Darwin argue, if there really is no good evidence for it?

'In sharp contrast with Darwin's ideas, the fossils tell us that the species arise in a mysteriously *abrupt way of sudden appearances,* after which come long periods of stability, followed, at times, by mass extinctions. Darwin said Nature does not make leaps. Yet, the repeated big and little bang appearances of new species, which the fossils actually show, tell us otherwise.

'So, those, like Darwin, who are materialistically inclined, can continue to try to explain these mysterious, sudden appearances of new species in terms of pure luck or unintelligent design (UD). Others, like Behe, Meyer and Dembski, can argue for subtle, intelligent causation, which they refer to as ID.'

'Ok, Ollie, but the problem for those, like Michael Behe, who argue for ID, is that they have no proposals as to *how* nature's designs, if they are intelligent, arise, who or what is responsible for them, or why.'

'I understand, Ali, this is a difficulty for them. Because, if they say, "We believe that the spiritual agency or agencies of our personal beliefs are responsible, be it instantly or by a guided evolution over time," the skeptics will say, "We can't see the creative god or angels or demons of your private beliefs. You are just appealing to unseen intelligences of the gaps."

'To which they can reply, "All your *materialist theories* of Life, including Darwin's, rely, just as much, on *'never-to-be-seen lucky-chemical-coincidences of the gaps.'* You too are relying on *unseen causes* which, mysteriously *'just are.'* After all, why do all the random, mindless chance events, which you believe to be so incredibly creative, *exist?* You, cannot say, other that they come from your MUA-UGM which *'just is.'* The only difference is that your unseen causes, which *'just are, who knows why?,'* are completely mindless and aimless, while ours, which also *'just are,'* are intelligent and purposeful.

'Further, we have demonstrated, not once, but repeatedly, that Life could *never* arise by mere chance, no matter how many 'lucky chances' you allow, and that there are, today, not just one or two but many valid scientific ways to show *positive* evidences for intelligence and design in living nature, even if the ultimate sources of those designs cannot be seen in physical terms. Can you, though, demonstrate that pure, dumb luck can create anything remotely intelligent or mechanically functional? No, it is is easy to show you cannot."

'They can also make it clear that their arguments for intelligent causation are not intended to, and cannot, point, directly, to the spiritual agent or agencies of their private beliefs but only to highlight the fact that our existences *cannot* be explained away anymore as mere chemical coincidences, as materialist thinkers like Darwin and Dawkins have sought to do.'

'So, Ollie, did you manage, in the end, to convince your friend, Vicki, in favor of intelligent causation versus pure chance as a creative possibility?'

'Maybe to some extent. But, for a lot of people, today, like her, to admit to the possibility of intelligent causation, although you might think it would be an interesting or even exciting hypothesis, would necessitate a major and potentially discombobulating shift in their thinking.'

'How so?'

'Well, if you concluded, long ago, as, for example, I did, when I came across Darwin's theory as a teen, that Darwin was right and that all talk of the supernatural was delusional, it can be difficult to change your mind.

'As a child you may have believed, in your own naive, childlike way, in the supernatural and the miraculous, that there were a god or gods, some loving or fearsome final force lying behind or beyond outer, visible reality.

'But, as you grew older, you began to question. You knew pain and loss and people were not always kind. You heard that prayer could be very powerful, so you prayed really hard for something you really wanted, but you did not get it. You were told, 'God is Love,' but people got ill and died. At times they slaughtered each other, in horrendous wars, and no one, you spoke to, had any very good answers to your questionings.

'They said that God had given us *free will,* so it was not His or Her or Their fault, (depending on how your culture conceived of these subtle forces), that people *misused* their free will. Ok, that might be true of people and their choices, but what about all the horrible diseases humanity was prey to? How could a good god or gods allow, let alone create such things?

'Then, you came across Darwin and a lot of things seemed to fall into place. What if, as he tended to believe, there were no – benign or malign – creative forces operating behind the scenes? What if the material world was all that existed and Life had arisen as just some freak, chemical accident and that, as he proposed, it evolved from there into all the amazing Life forms, including we ourselves, for no reason or point at all? *There was no meaning or intelligent spiritual creativity involved, it was all just pure, stupid, pointless luck.'*

'You might have felt a little sad that Darwin had swept away the naive spiritual beliefs of your childhood, but some things now seemed to make more sense.

'Perhaps you read some of Richard Dawkins writings and you agreed with him that, "In a [mindless] universe of . . . blind physical forces. . . , some people are going to get hurt, other people are going to get lucky, and you won't find any rhyme or reason in it, nor any justice. The universe that we observe has precisely the properties we should expect if there is, at bottom, no design, no purpose, no evil, no good, nothing but pitiless indifference.'[224]

'It *was* a harsh world, bullies often went unpunished, some people did terrible things in the names of their religions, which were supposed to be all about making the world a better place, and trying to do 'god's will.'

'Well, if this was religion, you wanted nothing to do with it. One religion, you were taught, was about 'love,' but so many atrocities had been committed in its name. Another, was said to be a religion of 'peace,' but was it? Another said, 'All are One' but taught people to believe in a rigid hierarchy of classes. So many contradictions. Darwin, you felt, provided an escape from all that.

'Religions might have some good points, but, on the whole, you began to think, we'd be better off without them. From this point on, you considered yourself a 'realist,' no longer superstitious, no longer needing to believe in a Father (or Mother) Christmas type God-in-the-Sky, who could make it all better – except he or she never did – and you got on with your life.

'Maybe, occasionally, out of habit or nostalgia, you went to your childhood church, temple or mosque but, apart from some cultural comfort and pleasing some of the more traditional members of your family, it didn't speak to you. Privately, you wondered, "How can people still believe this stuff?"'

'As to any modern – *non religious* – evidence for the supernatural, for psychic or paranormal phenomena of various kinds, it was of no interest to you. You'd made your decision to put your *faith* into materialism, and, to your mind now, any claimed evidence for the supernatural was just hallucination, delusion or downright fakery. No, for you, materialism equalled rationality.

'Can you imagine, Ali, for someone who has thought in these ways, perhaps for decades, how hard it might be for them to change their mind? To consider they may have been mistaken? It might not be easy. It would need a big shift.'

'I see what you mean, Ollie. But it's sad that so many people now seem to think that their only choice is between Darwin's materialist worldview or an over-naive, scripture-literalist religiosity which requires them to believe all kinds of things which do not make sense today or are positively harmful.'

'I agree, it is sad. I think the religions, themselves, have a lot to answer for here. It is sad that they are mostly no more interested than our current science is in any of the *plentiful modern research* into, and good quality evidence for, supernatural phenomena of various kinds. Evidence which can serve as *a useful bridge* between the scientific and religious worlds. No, generally their position seems to be that you must believe what they tell you on *'faith'* or you won't *'get it,'* or, when they are at their most negative, that you'll go to hell!'

---

[224] Dawkins, R. *River Out of Eden: A Darwinian View of Life.*

'You seem to be talking about a kind of 'third way' here, Ollie. A way between the current mainstream scientific view – which is so materialistic – and the old religious ways which seemed to depend, so much, *on faith alone.*'

'Yes, I think there is a need for a third way between: 'Science-as-an-unquestioning-faith-in-materialism' and religion as 'just-believe-what-we tell-you-and-don't-ask-too-many-questions!' A third way which properly allows for the plentiful, modern – *non-religious* – evidence for the supernatural.

'It is ironic, though, that it takes a huge and almost *unreasoning faith* to believe that mindless chemical particles of carbon, oxygen, hydrogen and other elements could combine, for no reason or purpose at all, to become you and me! Beings who can *feel and think and create* and do so many amazing things.

'Why don't people see the illogicality in that?' Wondered Alisha.

'Well, sadly, Ali, like my friend, Victoria, they just don't. To her it made little difference what I said. I could explain for as long as I wanted that Life could *never* arise by chance, that this was scientifically and mathematically proven, that even various very well known materialists admitted as much, but it made little difference. She didn't seem to be interested in the evidence. She simply blocked out what she didn't wish to know.

'To my mind, though, although she claimed to be 'scientific,' she was, like so many other people today, mistakenly *conflating science with materialism* and clinging, emotionally, to a set of beliefs which 'worked' for her.

'She wasn't interested in the *real-world evidence* refuting Darwin or in the modern arguments for intelligent design, nor in the extensive, modern – *non-religious* – evidence for the supernatural, including that for mind beyond the body, and for life after death.

'As the philosopher John Searle pointed out, for many, their *very firm faith* in materialism, which they call 'scientific,' is not so much motivated by a genuinely "independent conviction of the truth" as by a distrust of what, to them, seem to be the only alternatives. Namely the choice, as they see it, "between a 'scientific' approach, as represented by... 'materialism,' and an 'anti-scientific' approach, as represented by Cartesianism [dualism] or some other traditional religious conception of the mind." [225]

'So what can we do about that?' Wondered Alisha, looking a little glum.

'I guess all we can do is to point out that there is good *scientific* evidence now, not only for intelligent design but, also, for various supernatural phenomena, some of it verifiable and repeatable, that science and materialism are *not* the same thing, and that you don't have to be 'religious' to be open to ID or to other modern evidences for the supernatural.' We walked on.

---

[225] Searle, *The Rediscovery of the Mind,* 3–4.

22 Evolution As Religion

'In 1981, Colin Patterson, then senior paleontologist at the British Museum of Natural History, provoked an academic storm when he told an audience of fellow academics in America that evolution could no longer be regarded as knowledge but it had, like a religion, become a matter of *faith*. He said,

'Now, one of the reasons I started taking this anti-evolutionary view, well, let's call it non-evolutionary, was last year I had a sudden realization. ... One morning I woke up, and something had happened in the night, and it struck me that I had been working on this [evolution] stuff for twenty years, and there was *not one thing* I knew about it. That was quite a shock, to learn that one can be so misled for so long. ...

'So for the last few weeks, I've tried putting a simple question to various people and groups of people. The question is this: 'Can you tell me anything you know about evolution, any one thing that you think is true? I tried that question on the geology staff in the Field Museum of Natural History, and the only answer I got was silence. I tried it on the members of the Evolutionary Morphology Seminar in the University of Chicago ... and all I got there was silence for a long time, and then eventually one person said: *'Yes, I do know one thing. It ought not to be taught in high school.'* [226]

'Patterson continued,

'I shall take the text of my sermon from this book, Gillespie's *Charles Darwin and the Problem of Creation* ... He takes it for granted that a rationalist [i.e. materialist] view of nature has replaced an irrational one [a spiritualist view], and of course, I myself took that view, up until about eighteen months ago. And then I woke up and I realized that *all my life I had been duped into taking evolutionism as revealed truth* in some way. ...

'Now I think many people in this room would acknowledge that during the last few years, if you had thought about it at all, you've experienced a shift from evolution as knowledge to *evolution as faith*. I know that's true of me, and I think it's true of a good many of you in here.

---

[226] *Can You Tell Me Anything About Evolution?* Presentation by Colin Patterson at the American Museum of Natural History. Nov. 1981. Emphasis added. A CD and annotated transcription of the talk, including Patterson's interactions with luminaries like Niles Eldridge in the Q&A and lively discussion that followed the talk was, at one time, available here www.arn.org.

'So that's my first theme. That evolution and creationism seem to be showing remarkable parallels. They are increasingly hard to tell apart.' [227]

'So this was another world-class expert admitting to an audience of fellow academics that their discipline had become more like a religion than science?'

'Yes. You see, Ali, to *describe* the fossil evidence, in terms of matter and mechanism, in terms of Aristotle's first two causes, which has been done very thoroughly by now, *is not* the same as explaining it. Discovering that,

> 'This set of animals, with no obvious ancestors, *suddenly appeared, did NOT morph into new types of creatures,* as Darwin thought might happen, but stably remained the same, before, eventually, dying out. Then, other species mysteriously materialized in the same geologically abrupt way, stably remained the same before they too disappeared into extinction, only to be followed by further Life Waves of similar or different species . . .'

'doesn't tell us what *caused* those sudden – *non-evolutionary* – appearances of new species. Our modern knowledge belies the kind of young-earth creationism proposed by Bishop Ussher in the 1600s. But it doesn't show that *all* ideas of intelligent causation, what Aristotle referred to as formal and final causation, are false. And it gives no comfort to the self-contradictory Darwinian concept of a *random, purposeless, yet astonishingly progressive,* evolution culminating in the amazing abilities of humanity.'

'So it should not be taught in schools all over the world as *fact?*'

'No. To tell young people that millions of years ago many new creatures abruptly arrived in the Cambrian Big Bang followed by various further mysterious Life Waves and mass extinctions, including the story of the intriguing dinosaurs, is reasonable, because it's what the fossils show.

'To admit that we don't currently know, scientifically speaking, what caused the diverse species to arise in the first place is also reasonable. To concede that Darwin's theory does not accord with, nor does it explain the fossil evidence, *nor* the similarities in vertebrate limbs, *nor* molecular biology, *nor* embryology, *nor* homologies, *nor* genetics, *nor* the strange convergences and divergences between *marsupial and placental* mammals *nor* the fascinating convergence between the mammalian camera eye and the clever octopus's camera eye, *nor* the amazing, DNA CODED complexities of Living things, *nor* the astonishing caterpillar–to–butterfly metamorphosis, *nor* the biochemistry of vision, *nor* awareness, *nor* feeling *nor* thinking, is also reasonable. To admit that, currently, there are *no* conventional theories to explain any of these remarkable things that really work, would also be truthful.

---

[227] Ibid. Emphasis added.

'To explain to young people that *some* scientists are philosophical materialists – and they believe that all things are, essentially, cosmic accidents – is also fair because it's what they believe. Right or wrong, for them, all things in Nature are *unintelligently designed* and coincidental, right up to their own and everyone else's *intelligent* minds.

'The young people can also learn that, while science is a systematic and methodical search for truth, its practice can be affected, just as well known biologist Richard Lewontin stated, by scientists' prior commitments to what they already believe to be true or hope to be true.

'Those scientists, like Darwin, Dawkins and Lewontin, convinced that only mindless, solid-little-billiard-ball matter is real, will put forward purely materialistic theories. While it may seem counter intuitive to most of humanity, they train themselves to look at Nature, including their *own highly intelligent and highly purposeful existences,* as if it's something wholly unintelligent, with no meaning or purpose or design, and this hypothesis, which they find intellectually satisfying, can continue to be explored.

'Other scientists incline to philosophical idealism. They remain open to the classical idea that living things may have subtle, intelligent causes and that it may be possible, using modern scientific means, to demonstrate this. While those in this second group accept that many things in Nature are somewhat random, and 'unintelligently designed,' mountain ranges and river valleys, for example, they are skeptical that certain of its features, especially INFORMATION RICH, DNA CODED, *Living Nature,* let alone *intelligent minds* able to think and to create, to pursue science and philosophy, could fall together by pure, pointless, higgledy-piggledy, entropy-defying, chance.

'Yet, at the same time, it should be made clear to young people that modern scientific, ***non-religious,*** hypotheses of intelligent causation, such as those put forward by thinkers like Behe, Meyer and Dembksi, do not give credence to any particular set of religious, spiritual or political claims.

'In the same way, the modern ***non-religious*** evidence for mind beyond the body, for various kinds of *out of body* experiences and for life after death, which we will look at briefly in a while, does not, by itself, confirm the specific claims of any of the world's major religions except to affirm their ancient claim that physical 'death' is not the end of the *deeper* Life of the inner person, the *Soul-being* temporarily inhabiting a particular physical body.

'In other words,' said Ali, 'the world's religious believers, it spiritualists and its philosophical idealists believe that there is an *underlying* intelligence at work in living things and that its effects can be detected and described in scientific terms even if its ultimate source or sources cannot, currently, be captured, scientifically speaking?'

'Yes, and as evidence tending to support the modern intelligent design view, young people can learn, for example, of the cell membrane dilemma highlighted by Jacques Monod.[228] Although he was, himself, a materialist, he rightly wondered: *'How* could the cell's selectively permeable membrane, which, among its other functions, stops the DNA *inside* the cell from being destroyed by the chemical environment *outside* the cell, arise without the very DNA which CODES for it in the first place?' They could learn that this, 'Chicken or egg?,' problem is an example of irreducible complexity which points to there being far more to Life than mere chemical coincidence.

'To explain that various contemporary scientists – like Behe, Meyer and Dembski – have developed robust techniques for working out whether patterns and signals of various kinds are random or meaningful would also be true. That it is now genuinely possible to use modern methodologies to detect intelligent design versus the random design produced by purely natural laws.

'They could learn that the contemporary search for extra-terrestrial intelligence (SETI) is a branch of modern *intelligent design detection science* (IDDS) – as are archeology, (is that mound a natural feature or is it artificial?), forensics, (was the deceased pushed or did he fall?) and cryptography (is that gibberish or is it code?) – and that SETI is based on the analysis of radio signals from space for signs of meaningful non-randomness.

'Young people could discover that some scientists have now realized that if segments of Life's digital (DNA) Coding were being beamed towards towards our Earth in the form of a radio signal, the conclusion of intelligence and artificiality would be unavoidable.[229] Why? *Because of the highly specific, the physically indeterminate, the non random and the utterly purposeful complexity* of Life's DNA instructions:

TTAAGGCC ATAGGAGC CAACGGTT CCTTGTAC CAACGGTT
TATAGGCC AGGAATGC GGCCTATA GTACCCTT AGGAATGC

'They could learn that contemporary researchers, like shCherbak and Makukov, *having noticed various indicators of artificiality* in the CODE, such that nucleon sums are multiples of 037, that the stop codons act as zero in a decimal system, and all the three-digit decimals (111, 222, 333, 444, 555, 666, 777, 888, and 999) appear at least once in the code, 'which also looks like an intentional feature,' [230] write,

---

[228] Monod, J. 1972. *Chance and Necessity.* Collins. pp 134–135.

[229] The 'Wow! signal' of the terrestrial genetic code. *j.icarus.2013.02.17*

[230] Ibid.

'But it is hardly imaginable how a [purely] natural [material] process can drive mass distribution in abstract representations of the code where codons are decomposed into bases or contracted by redundancy. ... NO NATURAL PROCESS can drive mass distribution to produce the balance ... amino acids and syntactic signs that make up this balance are entirely ABSTRACT since they are produced by translation of a string read across codons. ... In total, not only the signal itself reveals intelligent-like features [but] ... taken together all these aspects point at [the] ARTIFICIAL nature of the patterns.'[231]

'In these ways, students of natural history can discover that some scientists hypothesize that the appearances of purposeful functioning, beauty and design we see all about us are not illusions but real. At the same time, they can hear the counterarguments put forward by popular *materialist thinkers* like Charles Darwin, Francis Crick and Richard Dawkins.'

'Ok,' said Alisha, 'but knowing that some scientists are *materialists* and others are *idealists or spiritualists* wouldn't provide young people with a grand new theory of Life would it?'

'No, but at least they'd learn that science isn't beholden to any particular belief system. Rather, it's about keeping an open mind and trying to find out what's *true*. The fact that we don't currently know, from a scientific point of view, how the purposeful functioning, the 24/7 intelligence visible in living things arises, nor how it gives rise, second by second, to the diverse plants and creatures we see all around us, doesn't mean there are no such processes.

'Why not give young people the facts: 'Life on Earth appeared billions of years ago, the big bang of the Cambrian explosion took place, trilobites suddenly appeared, fish suddenly appeared, spiders suddenly appeared, the dinosaurs came and went, birds and bats appeared and so on – and explain that there is more than one way to interpret the data. For example, when learning that scientists have now discovered that Living Cells are staggeringly complex, vastly more complex than Darwin and his steam era contemporaries imagined, and that, as this expert in cellular biology writes,

'We have always underestimated cells. ... The entire cell can be viewed as a factory that contains an *elaborate network of interlocking assembly lines, each of which is composed of a set of large **protein machines** . . .* Why do we call the large protein assemblies that underlie cell function protein machines? *Precisely because, **like machines invented by humans** to deal*

---

[231] *Evolution News* review of the "Wow!" Signal of Intelligent Design published in the *Planetary Science Journal Icarus*. www.evolutionnews.org. Text in square brackets and emphasis added.

efficiently with the macroscopic world, these protein assemblies contain highly coordinated moving parts.' [232]

'they can also learn that, while many scientists still believe that such complex and purposeful processes could arise by pure, stupid, chance, not all agree.

'They can discover that there are substantial intellectuals, – like Denton, Johnson, Behe, Battson, Latham, Dossey, Tour, Meyer, Axe, Gauger, Wells, Witt and Dembski – who doubt that these modern discoveries provide good evidence for 'purposeless' and 'unintelligent' processes. Modern day free thinkers who question the currently fashionable belief that there are just mindless laws + mindless chemical interactions + a random star, the Sun, and all Living Beings *mere coincidences* of these basic but pointless givens.

'In other words, Ollie, you think young people could cope with knowing that there are these two fundamentally opposed philosophies of existence – materialism or naturalism *versus* spiritualism or supernaturalism – and science is affected by the wish to see which is, ultimately, correct.'

'Yes, why not? We should explain that there are these *two basic hypotheses* of reality and we should not mislead them into thinking (a) that a purely physical science has all or the only valid answers, a delusion called *scientism*, (b) that it was, and it remains, scientifically established by Darwin, among others, that Life is just an accident which evolved by accident, nor (c) that science has proved materialism to be true.

'What Colin Patterson highlighted with his question: 'Can you tell me anything you know about evolution... that you think is true?,' is that Darwin's theory has, in our day, become a *faith* position, almost a cult in fact, because, in the intellectual West at least, Darwin's ideas are 'answers' to Life's mysteries which cannot be questioned, without academic risk, – even though the evidence very clearly falsifies them, not just once or twice but, repeatedly.

'Patterson was correct. No matter how often it has been pointed out that Darwin's once intriguing-seeming scheme cannot accommodate the data of the fossils, embryology, genetics, homology, molecular biology or the abrupt, the very non-evolutionary appearances of the species, the faithful take no notice.

'Darwinism has become a creation myth for materialists and for those who, it seems, prefer to have 'answers' in science which cannot be questioned over questions and mysteries which cannot, yet, be answered scientifically.

'One odd thing, though, Ali. If you ever listen to a discussion about how science is *supposed* to be, you'll be told how scientists always keep their minds open to new ideas and to new evidence. That, they have an ever ready

---

[232] Bruce Alberts. 'The cell as a collection of protein machines: Preparing the next generation of molecular biologists'. *Cell*, 92 (February 8, 1998): 291.

willingness to think again, to revise their earlier theories, and to humbly follow the evidence wherever it may lead. In many areas of science this may be true but, when it comes to current origin of life thinking and to Darwin's ideas, it seems not to be.

'We have seen, over and over again, how the data doesn't match Darwin's predictions. Yet, also over and over again, the data is ignored. We've seen how methodological materialism and methodological naturalism, although useful in places, can become science blockers, especially when used, or misused, to close down various perfectly reasonable lines of scientific inquiry.

'We have seen how brilliant modern Galileos, like Behe, Wells and Meyer, who put forward alternative ideas, are ridiculed or sidelined. Don't you find this odd? Scientists are supposed to keep an open mind. But, even when, so very clearly, Darwin's ideas cannot make full sense of Life and Mind on our planet, relatively few in the mainstream seem interested.

'So what's the answer?' Asked Alisha.

'Well, one answer is that our own *purposeful desires and our own capacities to do intelligent science* ought to tell us that *unintelligent chemical coincidences,* in however many warm little ponds, could not possibly lead to Life, let alone explain the arising of the very faculties in us which we call *mind and intelligence* and which permit us to do *rational, intelligent science!*

'It is only if we adopt *materialism* as our basic creed that, like Charles Darwin, and his modern super champion, Richard Dawkins, we will be forced to reject all notions of intelligence and purpose in Nature. But this takes our science away from an openminded investigation into reality – be it wholly physical or also spiritual – and turns it into a belief system.

'A belief system which makes us, and our science, meaningless. For, if we can call a Jaguar car or Jaguar fighter jet a cleverly designed object of beauty and purpose but, in the same, not so reasonable, breath, we are asked to pretend that its *living* feline inspirations, vastly more clever, more beautiful and more complex than *anything* we make, are merely the unintended byproducts of mindless chemical coincidence which only *'seem'* Soulful and purposeful and Alive, then our language and our science are no longer instruments for truth but misleading tools for reality-distorting lies and, whether we realize it or not, they have become positively *mephistophelian.*[233]

'It is only for those who reject or ignore all ideas of, and all evidences for, the spiritual and the supernatural, that some wholly materialistic scheme, of

---

[233] Goethe's Mephistopheles tricked Faust into selling his own Soul. The word 'could derive from the Hebrew *mephitz,* meaning "destroyer", and *tophel,* meaning "liar"... The name can also be a combination of three Greek words: "me" as a negation, "phos" meaning light, and "philis" meaning loving, making it mean "not-light-loving,"' wikipedia.

unintelligent design, like Darwin's, will 'have to be true', regardless of the data. Their prior philosophy, as Lewontin noted in his famous quote on materialism in science, will rule out any other possibility.

'But,' queried Alisha, 'it's difficult to understand why, if neither the fossils, molecular biology, nor embryology support Darwin, and if this was pointed out a long time ago, people still cling to his way of thinking?'

'Well, it's partly because as Chris Carter writes,

'In our modern world, science and scientists hold a great deal of prestige, and few people want to be thought of as unscientific. ... If to be scientific is good and unscientific bad, and if the term "scientific" is thought to be synonymous with the term "materialistic," then any talk of disembodied minds or spirits [or intelligent causation] is anti-materialist, unscientific, and therefore bad.[234] *The long-standing confusion of materialism with science* is what largely accounts for the persistent social taboo responsible for the ignorance and dismissal of the substantial amount of evidence that proves materialism false.'[235]

'But as Carter adds, 'the difference between science and ideology is not that they are based on different dogmas; rather, it is that scientific beliefs are *not* held as dogmas, but are open to testing and hence possible rejection.'[236] Unfortunately, Ali, materialism is not "open to testing and possible rejection."

'Why not?'

'Because, for its followers, it cannot be questioned. Its central dogmas:

'Only the *physical* worlds are real. *Mind and Feelings are illusions.* They are mere accidental byproducts of matter – mere epiphenomena of *mindless and feelingless* little-atomic-billiard-ball elements like carbon and oxygen.

'Life on Earth is just an astonishing accident, which evolved by accident, and Darwin proved it. Aimless chemical luck is supremely creative. All evidence for intelligence and design in Nature, and for the supernatural in general, no matter how convincing, is either self-delusion or fraud.'

'While these ideas are so weak and so wrong, on so many levels, as to be beyond unreasonable, and, frankly, absurd, everyone has a worldview, including scientists. This only becomes a problem when people become unwilling to consider evidence which may contradict their worldview.

---

[234] Carter: "For the materialist, the term unscientific seems to be the modern equivalent of the term heretical, and it is invoked for the same purpose: to exclude from consideration ideas that challenge the believer's faith."

[235] Carter, C. 2010. *Science and the Near Death Experience.* Inner Traditions; p.238.

[236] Ibid, p.237.

'For too long, Ali, we've been presented with a false choice by those who control science today. Either believe the world was instantly created a just few thousand years ago, by a ridiculously simplistic, Father Christmas type, god, which no rational person should believe in or needs to believe in, or, believe, not only that it's far more ancient but, that it isn't a product of intelligent causation at all. It's all entirely accidental and utterly without meaning – all the way up and all the way down!

'Patterson was right. Nothing about Darwinian *macro-evolution* is true. What some refer to as *micro-evolution,* finch beak and peppered moth variations, is different. It's not controversial and we can all accept it. But, it's now clear, this kind of relatively minor change could not and did not lead to one creature gradually transforming into another, as Darwin predicted should be the case if his theory be true.

'"Darwin said so, 150 years ago, so it must be right," is not evidence. Fanciful drawings of *imaginary* evolutionary trees and *imaginary* transitional creatures – which NEVER TO BE FOUND in the real world's fossil records – may impress some but they are *not* evidence.

'This is topsy-turvy. It is doing science backwards. Darwin had an interesting idea. But, if true, should it not be borne out by the real-world evidences of today's species, the fossils and genetics and embryology?

'Too many people, today, Ali, don't seem to notice the startling incongruence, the shocking dissonance, when Darwin's supporters say, "Finch beaks can change, 'Yes, that's true,' peppered moths can vary, 'Yes, that's also true,' there are many different varieties of dog, 'We agree,' and this proves that warm pond water struck by lightning could accidentally turn into *stunningly complex DNA CODED Living Cells."* 'What!?'

"Then, these 'accidental' single-cell microbes randomly morphed into worms, snails, spiders and fish." Really? How? Has anyone shown that this actually happened, let alone how it might be possible? "Water breathing fish, then, due to yet more random chemical changes, became air breathing amphibians," How?, "then dinosaurs and, eventually, brilliant minds like those of Plato, Beethoven and Einstein." An incredible scheme which is not only not supported by the data but is demonstrably impossible. It is extraordinary that all this is not seen more critically.

'Changes over time? Yes. The mysteriously abrupt emergences of new species? Yes. Similarities in structure and design? Yes. The process Darwin envisaged? No. An instant creation, 6000 years ago, based on Bishop Ussher's attempt to treat the Bible as if it was an early but accurate science book? No.

'All we really know, Ali, is that Life on Earth is extremely ancient and that it has gone through many changes. Beyond this we still know very little, scientifically speaking, about Life's true origins and what drove the mysterious, sudden appearances of new species, the long periods of stability, the disappearances, followed by fresh Life Waves.

'Our current science has no useful ideas as to what causes any Living Being to be, or Mind or Emotions to arise out of the seemingly mindless atomic dust of this world, this astonishing second and every astonishing second. No one has any scientific idea as to how *Living and Breathing, Sensing and Feeling, Feeding and Sleeping* all started or why, eventually, our physical bodies die.

'The mindless solid-little-billiard-ball atoms of the materialist belief system are not alive in the way we are alive nor do they need to breathe or to feed or to sleep, nor, even, do they die as we do.

'Yet our science is remarkably incurious about these things. It has nothing useful to say about them. As it denies the intrinsic intelligence, the subtle Life Forces and the Soul embodied in all living things, how can it?

'No one has any genuinely scientific idea why there are Cats with their feline natures and Dogs with their canine natures. According to materialism, they are just purposeless chemical coincidences. To those of us who are more spiritually inclined, they are *Living Beings* with their own unique states of consciousness and reasons for being.

'But as to precisely *how or why* they come to be no one can yet say. But it's very clear by now, if not so much in Darwin's day, that lucky chemical coincidences in no matter how many warm little ponds wouldn't do it.

'If our mainstream sciences could now just allow themselves to begin to admit that, currently, we do not have the beginning of a whisper of a real idea what Life and Mind *truly* are, nor what causes Living Beings to be *alive and sentient* in this world, we'd be on firmer ground.

'The current view, expressed by thinkers like Valtaoja,[237] is that it's all just accidental chemistry with no more Soul or intelligence to it than there is to a randomly evolving river valley and no more meaning or purpose to it than we ourselves care to make up.

'As to why, let alone how, the *Soulless, the Life-forceless, the Mindless* little billiard-ball-atoms of the materialist belief system would ever become Beetle or Eagle or Lion or People shaped or *Think* or *Feel or Soulfully care* about anything at all, let alone seek meanings or wish to pursue rational, intelligent, scientific, philosophical and spiritual inquiries, our current science, ruled by materialism, has nothing useful to say.'

We walked on by the lovely stream.

---

[237] Lesiola, M. 2018. *Heretic: One Scientist's Journey from Darwin to Design,* Discovery Institute Press. Valtaoja said to Leisola: *"Life is nothing else than physics and chemistry—mere electricity. There is no reason to assume anything supernatural."*

Part II

Beyond Materialism

Modern, Non-Religious Research *into* and

Evidence *for* the Supernatural

23 Spiritual Science

'In *Mind and Cosmos,* the distinguished philosopher, Thomas Nagel writes,

> 'If materialism cannot accommodate consciousness and other mind-related aspects of reality, then we must abandon a purely materialist understanding of nature in general, extending to biology, evolutionary theory, and cosmology.' [238]

'A science focused solely on matter and mechanism, Alisha, can describe many things, but it cannot fully explain them. To describe what a Life Form is *made of* and how it *works,* useful as it is, doesn't tell us how it arose in the first place. But the neglect of Aristotle's classical insight that all things have, not only material causes ('made of') and mechanical causes ('how its put together') but also, formal and final causes – shaping ideas and purposes – explains why we are in such an intellectual and metaphysical mess today.

'Materialism, where nature is conceived to be physical alone, has led our science to become so closed to the *Life and Soul* in things, not just in people and animals but also in Gaia, our amazing, living Earth. *Anima Mundi.* This is why our present science culture cannot allow itself to imagine that our earth may, in her own way, be a fascinating expression of *living, multihued Being, enSouled and enSpirited,* as the world's earlier peoples knew, some still know.

'No, today, we're asked to believe she's just a random rock. All her rivers and oceans, her forests and mountains, are supposedly dead and Soulless too, because, it is held, there are no Life Forces, [239] no Soul, no intelligence, no meaning to anything, just accidental chemicals having chemical accidents and, oh, so very ironically, highly intelligent and highly purposeful scientific minds trying to tell us just how utterly 'unintelligent' and 'purposeless' it all is.'

'For this limited, blinkered thinking, Gaia cannot be our immediate, Living source, a vast Consciousness in her own right, Mother Earth in a Soulful, *multi-dimensional* Cosmos of intelligence and meaning, Pachamama, as the Andean peoples call her. No, she's just a random rock, even though, as James

---

[238] Nagel, T. 2012. *Mind and Cosmos: Why the Materialist Neo-Darwinian Conception of Nature is Almost Certainly False.* Oxford University Press.

[239] Due to artificial synthesis of urea in 1828.

Lovelock explained,[240] like all Living organisms, she has many self-regulating and homeostatic features which make Life possible.'

'Sadly, it's not just materialists who reject this ancient view but some of the world's most well known religions, because, to them, it smacks of animism or nature worship. Regrettably, they seem to believe that the God or Spirit of their beliefs would give them a *dead* world, rather than a *Living World, an en-Souled World,* an enchanted, magical world, *Anima Mundi.*

'What happened to spiritual or divine *immanence, to the Living Soul and Spirit* in our understandings? Do these religions not see that it was the denial of the soulful and the spiritual *within* nature and the persecution of those they dismissed as animists and pagans, including many who we now think of as the world's first peoples, which laid the foundations for today's dismal spirit of materialism? An ideology which, finally, denies the *Soul and the Spirit* altogether.

'Some people think that if we acknowledge the *Soul and the Spirit* in the mountains and the rivers, in the woods and the trees, in the great oak and the giant sequoia – there long before we were born and long after we are gone – we are being fanciful or pagan. But, if *Soul and Spirit are All and In all,* then we are merely acknowledging the *inner living Beingness, the Soul in all things.*

'In his near death experience, after being struck by lightning and floating outside of his body, described in *Secrets of the Light,* Dannion Brinkley writes,

> 'As all this transpired, the vibrant kaleidoscope of living colors that emanated from Tommy and Sandy [trying to resuscitate him] astounded me. In fact, as I glanced around the room everything appeared to be literally alive and vibrating with color. Even the wooden chest of drawers in the corner radiated a multihued energy. What an amazing observation for a redneck! I wish everyone could see what I saw on that night. *It certainly changed the way I relate to animate, and even inanimate, life to this very day.* I no longer take for granted the unique and beautiful spiritual life force flowing through every creation in the physical world. To witness how intricately everything was connected, interwoven, and related, at the deepest levels of a highly organized matrix of networked energy, was indeed an overwhelming and inescapable new reality for me.'[241]

'As Dannion Brinkley found, the spiritual and the supernatural, the Life Forces and the Soul, are *all around,* not solely in some parallel dimension next

---

[240] Lovelock, J. 1979. *Gaia: A New Look at Life on Earth.* Oxford University Press. Lovelock, J. 2010. *The Vanishing Face of Gaia: A Final Warning.* Penguin

[241] Brinkley, D. & K. Brinkley. 2012. *Secrets of the Light: The incredible true story of one man's near-death experiences and the lessons he received from the other side.* Piatkus.

door. They are in the animals in the fields, in the children at play, in the wind waving leaves and the refreshing breeze. The forests and the mountains, the rivers and the seas are all alive in their own way.'

We walked on by the quietly flowing river, a peaceful herd of golden brown cattle grazing the lush meadow grasses, sheep further up the valley. After a while I said, *'Physical Science,* Ali, is based on physical measurements and, when circumstances permit, repeatable experiments. What we can refer to as *Spiritual Science,* on the other hand, of which the world's esoteric traditions, like yoga, shamanism, theosophy, anthroposophy and cabala, are parts, is based not only on the data of outer, physical reality but also on data obtained by *intuition and extra-sensory* perception, faculties latent in all of us, faculties we can develop if we wish...'

'Go on,' said Alisha, looking intrigued.

'Well, there have always been those who have been able to see, to some extent, into the *subtle dimensions of reality,* either by virtue of natural gifts or by inner training, such as some yogis and psychics undergo, or, indeed, anyone with sufficient interest and the right aptitudes can undergo. Rudolf Steiner, the famous Austrian philosopher, clairvoyant and spiritual-scientist, wrote,

> 'There slumber in every human being faculties by means of which he can acquire for himself a knowledge of higher worlds. Mystics, Gnostics, Theosophists – all speak of a world of soul and spirit which for them is just as real as the world we see with our physical eyes and touch with our physical hands. At every moment the listener may say to himself: that, of which they speak, I too can learn, *if I develop within myself certain powers which today still slumber within me.'* [242]

'Such people, advanced esotericists, yogis, mystics and psychics have seen into these deeper realities and reported their findings. Many otherwise everyday people have also had extraordinary psychic and mystical experiences, precognition, telepathy, near death and out of body experiences. This sort of evidence can be reviewed by both the original experiencer and by others. And this kind of research forms part of what, today, we can call *spiritual science.*

'For example, in the late nineteenth and early twentieth centuries, the theosophists, Annie Besant and Charles Leadbeater, directly explored the sub-atomic world using extrasensory perception or ESP. In his book, *The Secret Life of Nature,* Peter Tompkins describes how he set out to see if there 'really was an acceptable correspondence' between Besant and Leadbeater's description of atoms and 'the 'reality' of orthodox physicists.' He wrote,

---

[242] Steiner, R. 1918. *Knowledge of Higher Worlds and its Attainment.* Rudolf Steiner Press. 6th Ed. 2004.

'To find out I went in search of the first qualified theoretical physicist to reevaluate the theosophists' pioneering work in *Occult Chemistry*,[243] Dr Stephen M. Phillips, a professor of particle physics. Phillips's book *Extrasensory Perception of Quarks,* published in 1980, while dealing with the most advanced nuclear theories, including the nature of quarks, postulated particles even smaller than quarks as yet undiscovered by science. Analyzing twenty-two diagrams of the hundred or so chemical atoms described in *Occult Chemistry* . . . at the turn of the century, Phillips found it hard to avoid the conclusion that "Besant and Leadbeater did truly observe quarks using ESP some 70 years before physicists proposed their existence". What is more, their diagrams indicated ultimate physical particles even smaller than quarks.

By the time I discovered Phillips on the southern coast of England . . . he had checked another eighty-four of the theosophists' atoms: all were seen by him to be 100 percent consonant with the most recent findings of particle physicists. Every one of the 3,546 subquarks counted by Leadbeater in the element of gold could be correctly accounted for by Phillips. Were Phillips's conclusions to be substantiated by his peers, it would adduce evidence that the theosophists with their yogi powers had effectively opened a window from the world of matter into the world of spirit.

Prompt and committed approval of Phillips's conclusions had already come from the noted biochemist and fellow of the Royal Society, E. Lester Smith, discoverer of vitamin B-12. At home in both the mathematical language of physics and the arcane language of theosophy, Smith spelled out his support in a small volume, *Occult Chemistry Re-evaluated* [1982]. And Professor Brian Josephson of Cambridge University, a Nobel Prize winner in physics, was sufficiently impressed by Phillips's radical thesis to invite him to lecture on the subject at the famous Cavendish laboratory in 1985.

Yet few in the ranks of orthodoxy had the courage to risk their positions by supporting anything so wild as the notion that psychics could see better into the basic constituents of matter than could physicists armed with billion-dollar supercolliders.' [244]

'That's intriguing, Ollie.'
'Yes, and it's very important.'
'Why?'

---

[243] Besant, A. Leadbeater, C.W. C. Jinarajadasa. 1908. *Occult Chemistry.* Theosophical Publishing House

[244] Tompkins, P. 1997. *The Secret Life of Nature,* Thorsons, pp. 68–69.

'Because science depends on good quality evidence, on *sensory* evidence. According to the materialist view, we only have physical senses. But if there are *other, subtle* senses which we can use to gain reliable knowledge not only of the supernatural but also of mundane nature, then *the materialist paradigm is refuted.* Leadbeater and Besant used such *other* senses. They falsified materialism. The implications of their work should have been epoch making.'

'Okay,' said Alisha, 'if Leadbeater and Besant used psychic perception to accurately describe sub-atomic particles, some time before conventional scientists even knew of their existence, why isn't this more widely known?'

'Because, when their book, *Occult Chemistry,* came out, a hundred years or so ago, no physical scientists had seen inside an atom. Therefore, they had no way to evaluate the theosophists' fascinating descriptions. So it wasn't until, much later, a professor of particle physics, Dr. Stephen Phillips, came across a copy of *Occult Chemistry* that their accounts of the particles could be verified. Peter Tompkins writes,

> 'Phillips counted the quarks, and saw that the Theosophists had the correct number of quarks in every element, and the last quark was only discovered in 1997 – the proof of it is self-evident. . . .
>
> 'The excuses for disbelieving the claims of the psychics are irrelevant in the context of their highly evidential descriptions of subatomic particles published in 1908, two years before Rutherford's experiments confirmed the nuclear model of the atom, five years before Bohr presented his theory of the hydrogen atom, 24 years before Chadwick discovered the neutron and Heisenberg proposed that it is a constituent of the atomic nuclei, 56 years before Gell-Mann and Zweig theorized about quarks. Their observations are still being confirmed by discoveries of science many years later.'

'Professor Phillips himself published a book, *Extra–Sensory Perception of Quarks* [245] confirming his findings but, according to Tompkins, although it was very interesting, it was,

> 'too scholarly – too much mathematics and physics in it for the general public. For [many] scientists themselves, it shatters their whole [materialistic and anti-supernaturalist] premise and if they don't want to look at something, they won't. In fact they will do their best to put you down . . .'

'In an interview, Tompkins was asked whether the fact that modern physics has confirmed the clairvoyants' descriptions of sub-atomic matter means that their extra-sensory descriptions of *the spiritual hierarchies,* the worlds of the angels, nature spirits, devas and fairies can also be taken as real? He replied:

---

[245] Philips, S.M. 1980. *Extra–Sensory Perception of Quarks.* Theosophical Publishing.

'If a hundred years ago the Theosophists were absolutely correct in the number of quarks which they described, then one has perforce to look at their detailed descriptions of nature spirits. Obviously they are using effective clairvoyance. When you find that the shamans all throughout South America, for instance, describe the same phenomena as the shamans in the Far East, rationally you must take it into consideration as being possibly real. Why would people accept the physicists' description of an atom, which the general public cannot see, and not accept the description of the nature spirits that they cannot see? They accept the religious notions of spirit [and spirits], though they cannot see them, why not accept the possibility of nature spirits and angelic hierarchy? . . .

'The whole Christian persecution of witchcraft, and the [denial of the] whole world of nature spirits, is what has alienated us from a healthy planet. It's only when we get back in touch and accept their presence that the whole thing falls back into place and we take our place in the cosmos. Otherwise we're totally alienated from a healthy planet and from our role in the cosmos.' [246]

'You see, Ali, the world's earlier peoples realized that there is no atom or crystal, no flower, tree, animal or person, no river, forest or mountain, no planet, star or galaxy of holistic reality which does not have *subtle, spiritual dimensions*. Tompkins and Bird's bestselling book, *The Secret Life of Plants* [247] showed how the Plant world is *pervaded* by intelligence and subjectivity. They described the amazing sensitivities of plants, their capacities to be lie detectors, their fascinating abilities to help each other and to adapt to human wishes, their responses to different kinds of music and their ability to communicate with humanity.

'Other thinkers, such as the formidable Blavatsky, founder of theosophy, and the titanic Rudolf Steiner, [248] founder of anthroposophy, biodynamic agriculture, Waldorf education, anthroposophical medicine and eurythmy, have also averred that all dimensions of reality – even the mineral – have subtle, sentient elements, as the world's first peoples always said.

---

[246] *Kindred Spirit Magazine*. 1998, June.

[247] Tompkins, P and C. Bird. 1973. *The Secret Life of Plants*. Harper Perennial

[248] Rudolf Steiner: "Steiner's gift to the world was a moral and meditative way to objective vision... If accepted... it could bridge the existing cleft between a man's religious conviction and his intellect and will." Franz Winkler, M.D., *Man the Bridge between Two Worlds*. "That the academic world has managed to dismiss Steiner's works as inconsequential and irrelevant, is one of the intellectual wonders of the twentieth century. Anyone who is willing to study those vast works with an open mind (let us say, a hundred of his titles) will find himself faced with one of the greatest thinkers of all time, whose grasp of the modern sciences is equaled only by his profound learning in the ancient ones." Russell W. Davenport, *The Dignity of Man.'* Amazon.

'These people all take a *panentheistic* view of reality. Namely, that Self-Existing Intelligent Beingness, to which humanity gives its traditional religious labels, like God, Buddha Nature or the Tao, is not just out *there* but is, also, right *here*. It is not separate from the Eons-Long-Creation-in-Evolution (ELCIE) but it is both the outer *form* and the inner *Life* of everything. That every last quark, electron and photon of total holistic reality is a meaningful vector *for* and vibration *of* Spirit. Atoms vibrate at 10,000,000,000,000 cycles per second – that is very *alive!*'

'Are you saying the whole universe is *alive?*'

'In a sense, yes. Multi-dimensional reality consists, ultimately, in *Spirit*. It is composed, as India's yogi-adepts have always said, of *Creative–Being–Consciousness*. This is why they say, *'tat tvam asi' or 'That' you are.'* Which is to say, *'You, too, are woven from the very stuff of self-existent, 'I AM THAT I AM,' Beingness or True Nature.* You are made in the image of God or Being.'

Alisha spoke. She said quietly, 'This intuition that all is *alive, all is essence, presence and consciousness,* even down to the plants, minerals and crystals, is why it's so important to treat *all* of creation with as much respect as possible, not just grab, stake, take, make and break, but rather to say please and thank you, as the world's earlier peoples did and some still do.'

'Yes, that's very true. Listen, to what Steve Taylor says about this in, his fascinating book, *Waking from Sleep.*

'. . . to indigenous peoples there are *no such things as "inanimate objects"*. . . . they sensed that even rocks, the soil and rivers had a being or consciousness of their own. Whereas European or Asian peoples tend to see these things as nothing more than one-dimensional objects which they are entitled to use for their own devices, indigenous peoples *felt* that the natural phenomena around them were alive and so treated them with respect.. . .

'As anthropologist Robert Lawler notes, "Every distinguishable energy, form or substance has both an *objective* and a *subjective* expression." Aborigines have complained that the problem with European Australians is that they can only perceive the surface reality of the world and can't enter its interior life. As one aboriginal elder stated,

"Unless white man learns to enter the dreaming of the countryside, the plants, and animals before he uses or eats them, he will become sick and insane and destroy himself."' [249]

'Unlike most of us, Ali, those whom we now think of as the world's first peoples, the yogis and the mages, the shamans and the sages, realized long ago

[249] Taylor, S. 2010. *Waking From Sleep.* Hay House, p. 44.

that holistic Nature is both a unity, or non-dual, yet, at the same time, like a coin, it can be experienced as a duality, consisting in Subjects (Heads) and objects (tails), Mind and matter, Life and form, Souls and mechanisms, Spiritual and physical, Intelligent information (Life's Codes) and chemical carriers of information (deoxyribonucleic acid ink).

'They realized that we are not only *part* of the 'all that is' but we are *made* of it and that, as microcosms of the macrocosm, we are blessed with the magical possibility to *know* it. Originally, science began because people, noticing the orderly and predictable movements of the planets and the stars and the wisdoms embodied in the plant, animal and kingdoms, assumed that these had intelligent, not mindless sources. After Darwin, however, science became an activity in which *purposeful, intelligent* people, scientists, devoted themselves to explaining just how 'unintelligent' and 'purposeless' Nature is – without noticing the profound contradictions in their positions.'

'But the subtle causes of reality are not readily visible to most of us, so many doubt that they exist.' Commented Alisha.

'Yes. But we can deduce some of these subtleties and almost everyone has at least one or two spiritual experiences which point to deeper realities. Look, in the end, there are just two ways to determine whether intelligent causation plays a part in the creation of reality or whether all is chance.'

'Firstly, deductive logic and the laws of probability, which tell us that living things could not arise by chance, it's not merely highly unlikely, it's impossible. Secondly, our intuitions, including, sometimes, extra-sensory experiences, which tell us that Living things *do not look* unintelligent or accidental because they really are not. That when we look at a new born baby, or at the soft fresh buds on the trees in spring, we are not looking at irrational chemical accidents but at *miracles* literally unfolding before our eyes.

'Lesser miracles within the Greater Miracle which is to weave the Sacred Tapestry of Existence with Self-Expressed cosmic silk and wool, from no-thing to some-thing, from no-where to now-here. A view endorsed by the world's prophets, psychics and mystics together with that of many otherwise perfectly ordinary people who confirm these things.'

'Some people, Ali, have always known the universe is not only material but also spiritual. In fact, as many physicists now realize, it is not, ultimately, 'material' at all.[250] Rather, it is in-form-ed (brought into form) via high vibrating energies, which, ultimately, disappear into f...o...r...m.. l...e...s...s...

---

[250] Capra, F. 1992. *The Tao of Physics.* Flamingo. 'The... classic exploration of the connections between Eastern mysticism and modern physics. An international bestseller, the book's central thesis, that the mystical traditions of the East constitute a coherent philosophical framework within which the most advanced Western theories of the physical world can be accommodated, has not only withstood the test of time but is ever more emphatically endorsed by ongoing experimentation and research.'

n..e..s..s.., a mysterious shimmering smile of a Divine Dream full of magic and meaning.

'Bring physical-science and spiritual-science together and we arrive at a more complete understanding of nature's two sides: *outer,* consisting in matter and mechanism, from atoms to galaxies, and *subjective, inner, spiritual nature.*

'All of it an expression of *Creative–Spirit–Intelligence,* hologramatically self-fragmented into trillions of lesser sparks, diamonds and flames of consciousness, expressing as all the *subjectivities,* from the tiniest spirit beings, to the plants, animals, humans and angelic hierarchies beyond.'

'So, you're saying that modern physical-science can tell us a lot about Aristotle's first two causes, matter and mechanism ('made of' and 'how it's put together'), but little or nothing, currently, of the *subtle, subjective sides, the inner meanings* of the phenomena of the mineral, vegetable, animal and human kingdoms, the *Soul* sides of things. Whereas, the world's spiritual research traditions[251] can form part of an emerging *spiritual-science* and provide insights into these things?'

'Yes. For example, traditional religious creation stories along the lines of, "God said, 'Let the Worlds exist,' and they did," reflect humanity's ancient intuition that existence is meaningful. But they do not explain the *physical* mechanics of Life's workings. This was why, in the early 1600s, Francis Bacon, set out, in the *Novum Organum Scientiarum,* his epoch defining program for the development of *modern, physical* science.

'One of his insights was that, for modern science to take off, natural philosophers, as scientists were then called, should stop focusing, in a medieval kind of way, soulful as it was, on the ultimate meanings and purposes of things, their Aristotelian formal and final causes, but, primarily, on *what they were made of and how they worked, that is on materials and mechanisms.*'

'You mean they should practice methodological materialism and methodological naturalism or MM/MN?' Said Alisha.

'Yes, they should focus on physical observations combined with practical experimentation, rather than on ancient dogmas of how things 'ought' to be, based on old, often untested, theories. Later, they could always go back to exploring the formal and final causes of things – their meanings and purposes, a process based partly on deductive logic and, at times, more subtle ways of knowing. Bacon's program was an outstanding success. Partly thanks to him, the scientific revolution took off and we have all the benefits we enjoy today.

'We are, however, now coming to that exciting point when our sciences can start thinking again about formal and final causes, the meaning and intelligence

---

[251] Like shamanism, theosophy, anthroposophy, kabala, sufism, yogism and taoism

of things, and of *who and what we truly are*. We cannot do this by insisting, from the outset that, 'Only mindless, dumb matter is truly real and all ideas of spiritual causation are fanciful and impossible,' or that, 'We can, in any case, explain everything in purely materialistic terms *even if* the spiritual *is* real.'

'If we do that we are no better than the medievals who held things 'had' to be a certain way to fit their prior beliefs. The 'Sun had to orbit the Earth' even if it was not true. Today it tends to be, "Darwin *'must be'* right and materialism *'must be'* true even if these ideas turn out to be false."

'Science is supposed to be an open-minded quest for truth. So, was Darwin right? Did Life arise and evolve as he said? Or, is there more to reality than just physical stuff? Do out of body experiences take place or not? Is there any good evidence for them? Are afterlife communications possible or not? Is there any modern evidence for them? Is telepathy possible? Is there any evidence for it. Are these not interesting questions?

'As the philosopher Thomas Nagel says, if *materialism* cannot account *fully* for the data of reality, including, especially, for *Life, Consciousness, Feelings and Mind,* should we not question it? What if humanity's ancient insight that we are not merely beings of matter but also of *Soul and Spirit* is correct?

'What if the Nature from which we emerge is, as the world's first peoples intuited, *alive and ensouled?* [252] What if we are, after all, *multi-dimensional and multi-sensory* beings, not solely beings of physical senses and dry thought, but also of *intuition and higher sensory capacities* which permit alternative ways of data gathering and making sense of reality?'

'As Besant and Leadbeater demonstrated?' said Alisha.

'Yes. Instead of reflexively dismissing all such ideas, in an, 'Oh, aren't we so very clever?,' urbanite kind of way, why not begin to reacquaint ourselves with the deeper dimensions of things, with nature's *more subtle* elements? Like the subtle anatomies of the human *chakra* system and the energy *meridians* long known to the Indian and Chinese civilizations.'

We walked on in the gentle valley as the shadows began to lengthen.

---

[252] Goswami, A. 1993. *The Self-Aware Universe: How Consciousness Creates the Material World.* Jeremy P Tarcher.

24 Mind Beyond Matter

'Our current science is not yet much interested, Alisha, but there are various perfectly valid and reasonable ways to falsify the materialist thinking which presently rules. One way, as we have seen, is to show that living things really could not 'fall together' by mere chance and that while purely physicalistic theories can adequately describe the random formation of landscapes they are hopelessly inadequate to account for the purposeful formation of *living* things.

'Another way is to research modern, non-religious claims for phenomena like telepathy, psychokinesis, Life after death, spiritual healing, near death and other out of body experiences. There are reliable ways to put these claims to the test.

'A contemporary researcher is Rupert Sheldrake. He's a well known biologist willing to study phenomena which do not readily mesh with purely materialistic interpretations of reality. He has written about the results of his researches in *A New Science of Life*, *Dogs that Know When Their Owners are Coming Home*, *The Sense of Being Stared At* and *The Science Delusion*.

'By contrast, Richard Dawkins, also well known, is a conviction materialist who has actively campaigned against religion and belief in *supernatural* phenomena which materialism cannot explain. He is author of *The Selfish Gene*, *The Blind Watchmaker* and *The God Delusion* among others.

'Intriguingly, in 2007, he visited Sheldrake to discuss the latter's research on unexplained abilities in animals and people. The interview was intended to form part of a TV series called *Enemies of Reason* which Dawkins was making as a follow up to his criticism of religion called *The Root of All Evil*.

'Before *Enemies of Reason* was filmed, the TV company assured Sheldrake – who was cautious about taking part – that the documentary would be "an entirely more balanced affair than *The Root of All Evil* was." And that, "We are very keen for it to be a discussion between two scientists, about scientific modes of enquiry.' So, they agreed to meet. Sheldrake writes at his website that

'I was still not sure what to expect. Was Richard Dawkins going to be dogmatic, with *a mental firewall* that blocked out any evidence that went against his beliefs? Or would he be open-minded, and fun to talk to? . . .

'Richard began by saying that he thought we probably agreed about many things, "But what worries me about you [Rupert] is that you are prepared to believe almost anything. Science should be based on the minimum number of beliefs." I agreed that we had a lot in common, "But what worries me about you [Richard] is that you come across as *dogmatic,* giving people a bad impression of science."

'Dawkins then said that he'd like to believe in telepathy but there wasn't any evidence for it. He also put it to Sheldrake that it can't be possible because it would "turn the laws of physics upside down." A view which, as we saw earlier, rests on outdated ideas about physics. Dawkins also put it to Sheldrake that "Extraordinary claims require extraordinary evidence." Sheldrake replied:

'This depends on what you regard as extraordinary. Most people say they have experienced telepathy, especially in connection with telephone calls. In that sense, telepathy is ordinary. The claim that most people are deluded about their own experience is extraordinary. Where is the extraordinary evidence for that?'

'Dawkins responded that people are prone to wishful thinking when it comes to the *super* natural and the paranormal. Sheldrake continues,

'We then agreed that [1] *controlled experiments were necessary*. I said that this was why [2] I had actually been doing such experiments, including tests to find out if people really could tell who was calling them on the telephone when the caller was selected at random. [3] *The results were far above the chance level.* The previous week I had sent Richard copies of some of my papers, published in peer-reviewed journals, so that he could look at the data. Richard seemed uneasy and said, *"I don't want to discuss evidence."* "Why not?" I asked. "There isn't time. It's too complicated. And that's not what this program is about." The camera stopped.

'The Director, Russell Barnes, confirmed that *he too was not interested in evidence*. The film he was making was another Dawkins polemic. I said to Russell, "If you're treating telepathy as an irrational belief, surely evidence about whether it exists or not is essential for the discussion. If telepathy occurs, it's not irrational to believe in it." . . .

'Richard Dawkins has long proclaimed his conviction that "The paranormal is bunk. Those who try to sell it to us are fakes and charlatans". ... But does his crusade really promote "the public understanding of science," of which he is [was, at the time,] the professor at Oxford? *Should science be a vehicle of prejudice, a kind of fundamentalist belief-system? Or should it be a method of enquiry into the unknown?* '253

'That's interesting, Ollie, two scientists with very different attitudes...'

---

253 Rupert Sheldrake's Website: Dialogues and Controversies; emphasis and text in square brackets added.

'Yes, the two very different approaches to scientific inquiry we have been discussing. The first assumes, as a matter of methodology (MM/MN) and often, also, of philosophy (PM/PN),[254] that reality is only *material,* that it can be studied as such, and that all things have *purely physical* explanations.

'The second, Sheldrake's approach, is to remain open to a wider range of possibilities, including the classical hypothesis that there may be more to Life than mindless matter + mindless physical laws + mindless chance.

'Another scientist, prepared to consider a broader view, Harvard academic, Dr Gary Schwartz, Ph.D, author of *The Afterlife Experiments: Breakthrough Scientific Evidence of Life After Death,* 'asked some of the most prominent [spiritual] mediums in America to become part of a series of experiments to prove, or disprove, the existence of an afterlife. ... [Leading to] a breakthrough scientific achievement: contact with the beyond under ***controlled laboratory conditions.***

> 'In stringently monitored experiments, leading mediums attempted to contact dead friends and relatives of 'sitters' who were *masked from view and never spoke, depriving the mediums of any cues.* The messages that came through stunned sitters and researchers alike. There are some extraordinary and uncanny revelations, and ... Dr. Schwartz was forced by the overwhelmingly positive data to abandon his skepticism.'[255]

'After the experiments, Dr. Schwartz wrote,

> 'I can no longer ignore the data [on life after death] and dismiss the words [coming through mediums]. They are as real as the sun, the trees, and our television sets, which seem to pull pictures out of the air.'

'But aren't these things more in the realms of religion than science?'

'There's no need for such a restriction, Ali. While religions have their particular convictions about such things, others can conduct *modern* research, in a more detached way and with a more open mind. Australian lawyer and afterlife researcher, Victor Zammit, makes this point. He writes,

> ***"The objective [modern] evidence for the after life has nothing to do with religion or personal belief."*** [256]

---

[254] MM/MN being methodological materialism and methodological naturalism, PM/PN being philosophical materialism and philosophical naturalism. PM/N is materialism/naturalism not just as scientific method but as belief and ideology, where only the visible, physical universe is held to be truly real.

[255] Schwartz, G.E. Phd. W.L. Simon. 2002. *The Afterlife Experiments: Breakthrough Scientific Evidence of Life After Death.* Atria Books.

[256] Zammit, V. W. 2013. *A Lawyer Presents Evidence for the Afterlife.* White Crow Books.

'So, why is there a lack of mainstream interest in this kind of research?
'Because, as Chris Carter says, it is the,

'implicit equation of materialism with science, that explains the widespread practice of ignoring and dismissing the objectionable evidence as somehow 'unscientific.'

'Materialism is upheld as an incontestable dogma on which, it is thought, rests the entire edifice of science. But the difference between science and ideology is not that they are based on different dogmas; rather, it is that *scientific beliefs are not held as dogmas, but are open to testing and hence possible rejection*.

'Science cannot be an objective process of discovery if it is wedded to a metaphysical belief [like materialism or darwinism] that is accepted without question and that leads to the exclusion of certain lines of evidence on the grounds that these lines of evidence contradict the metaphysical belief.' [257]

'Can you put that more simply?'

'What he's saying is that *science cannot be objective if it simply assumes materialism to be true* – as thinkers like Darwin, Dawkins and Lewontin do – when this is the very question at issue.

'Is materialism true? Has it been shown to be? Could Life arise by pure chance? Could *bright, intelligent minds* accidentally emerge, from mindless matter, if you just allow enough time? These are the very questions we have been considering all through our long walk.

'One of the consistent challenges to materialism is the steadily accumulating *evidence for mind beyond the body*. There are, by now, many modern accounts of near death and other out of body experiences documented by bodies like the International Association for Near Death Studies and the Monroe Institute in Virginia. There are also many books, like Dr Raymond Moody's *Life After Life,* Margot Grey's *Return Form Death,* and the Fenwicks' *The Truth in the Light: An Investigation of over 300 Near Death Experiences.*

'In *Science and the Near Death Experience: How Consciousness Survives Death,* Chris Carter examines both ancient and modern accounts of NDEs from around the world, including the West, India and China. A reviewer writes,

'*Science and the Near-Death Experience* ... [is] "the best book on NDEs in years." Its strength is (among others) that *it takes all 'skeptical'*

---

[257] Carter, C. 2010. *Science and the Near Death Experience.* Inner Traditions; p.237.

explanations seriously, examines them thoroughly and demonstrates why they all fail. The necessary conclusion is that NDE/OBEs are, in fact, what they always seemed to be, and what all experiencers hold them to be: Experiences of a mind, which has left its body.

'Perhaps you will insist that "science" cannot accept this conclusion, because it must adhere to a materialistic monism? [only matter is real] Well, *what is the task of science? Is it to explain phenomena or to explain them away, following a pre-set ontology?* Is the ontological basis for science testable or not? If it is not, that basis is in effect a dogma. And science is no longer science.' [258]

'What does he mean: 'Is the ontological basis for science testable or not?'
'As Chris Carter pointed out, if scientists base their ideas on a particular set of *beliefs* about reality, like *materialism or Darwinism,* then for their claims to be scientific, their beliefs must be *testable*. Otherwise they're just ideology. Here are some conflicting ontological bases which can be tested.

(1) The Cosmos is *material* alone. No, it can be shown to be more than this.

(2) Mind *cannot* influence matter. Yes, it can, and this can be shown.

(3) Scientists can only detect *'unintelligent design'* in Nature. No, they are mentally *and* methodologically equipped to detect both unintelligent design (random events) and intelligent design (meaningful patterns).

(4) Life is just an accident. No, this is impossible and it can be shown to be.

(5) Mind cannot exist beyond the body-brain. It can do so and this can be shown.

(6) Life after death is not possible. It *is* possible and there is good evidence for it.

'The answers to these kinds of ontological questions will determine the types of inquiries scientists carry out. For example, if they decide reality is material alone, not only will they not be interested in investigating any of its *super* natural elements, but they may actively seek, like Richard Dawkins, to discredit any claims about them. On the other hand, if, like Rupert Sheldrake or Alfred Russell Wallace, they conclude that mind may be a principle in its own right, they will be more likely to conduct such research.

'In *To Heaven and Back: A Doctor's Extraordinary Account of Her Death, Heaven, Angels, and Life Again: A True Story,* scientist and surgeon, Dr Mary Neal, describes her near-death experience. In 1999, in the Los Rios region of southern Chile, while descending a waterfall, her kayak became pinned at the

---

[258] Carter, C. 2010. *Science and the Near Death Experience.* Inner Traditions, amazon, T. Skaftnesmo, text in square brackets added.

bottom. She experienced dying by drowning and left her body before being revived a short while later. In a subsequent interview she reported:

> 'I was acutely aware of everything that was happening. I knew that my efforts to exit the boat were not working, that I was out of air, and that I was too far from the riverbank for anyone to reach me. I knew that I would probably die. Having grown up with a fear of drowning, I was surprised to find my transition from life to death was seamless, peaceful, and beautiful. I felt quite wonderful. Before my near-death-experience, I believed in God and took my kids to Sunday school but was not particularly religious. Like many accomplished young adults, I felt like I was in control of my life and my future. Although I tried to be a "good" and "moral" person, my faith was not integrated into my daily life and the demands of work and family left little time to think about spirituality.' [259]

'Neal describes herself as *'a scientist by training, a skeptic by nature, and a very concrete, rational thinker.'* Yet, as for many other NDErs, the spiritual impact of her experience was major. Before, she was not particularly religious. Afterwards, her spirituality became a key part of her life. When asked why she thought she'd come back, she replied, as NDErs often do, that she didn't *want* to return but felt she had little choice, feeling she needed to share her experience with others and to support them in their faith. When asked what she'd like people to know about the afterlife she says simply:

> 'God's unconditional love for each of us is intense, complete ... Before we return to Heaven, our real home, we have an incredible opportunity on Earth to face challenges that will help us learn, grow and to become more Christ-like . . . Our time is so short that we need to be about God's business every day.' [260]

'Here is another remarkable NDE quoted by Dr. Bradley Nelson in his best-selling book on emotional healing, *The Emotion Code*. Anne Horne from Seattle wrote to him as follows:

> 'I originally came into contact with your [healing] work when a practitioner... used your techniques to balance my body... [and, at the end of the session, she] said, 'Oh, let's see if you have a Heart Wall...' She explained how emotions can put up a wall between yourself and others around your heart. [And how] she

---

[259] Neal, M. 2012. *To Heaven and Back: A Doctor's Extraordinary Account of Her Death, Heaven, Angels, and Life Again: A True Story.* Waterbrook Press. amazon interview.

[260] Ibid.

would run a magnet down a person's back to release those emotions and open their hearts.

'I couldn't believe what I was hearing. It was like an electric bolt went through me. Suddenly, an event that happened to me 25 years ago made sense. . . . When I was 23 . . . I had one of those near-death experiences . . .

It was a very important experience for me. I left my body and had a life review. . . . I was going back, home, and on my way, there was a light, a tunnel. I felt like I was being pulled by my heart toward a wonderful place. In that moment, I was encompassed by all this innate intelligence and tremendous love. And I just wanted to go home. It was fantastic.

'Anne was then told, to her disappointment, that it was 'not her time' but she was shown something in her future, back on Earth, where she saw people,

'doing a specific training that was very unusual. There would be 2 or 3 people together, with one person lying on a table, or standing up, and another person who was rolling something down the other person's back.

I knew this was in my future, that I was one of these people. I could feel the sense of urgency that they were feeling. It was like a numbers game; we had to treat as many people as possible. We were really in a rush, really hurrying. It was very, very, very vital. I couldn't quite understand what was going on, [so I asked him] 'what are we doing?' 'You are opening people's hearts,' he said. 'Not in a physical way. You're removing all blocks from their hearts so that they can give love and receive love from here.'

At that moment, the people doing this work became consciously aware of each other. It wasn't something planned . . . It just happened. We became conscious of each other. And at that moment, the meaning of this work became clear to me.

Suddenly a flood of energy was sent to the earth from where I was, above the earth. It looked like a white bolt of energy that came in through... our hearts into the world. We were there opening people's hearts so that they could be anchors for this divine energy to come into this world. Within three seconds, the world was completely transformed by this energy. This light went into every crack and crevice, everywhere, and there was no darkness in this world, ever again. The next thing that happened was, the doctors resuscitated me and I was brought back to life.'

'During my near-death experience, I said to the man who met me, 'But there are only thousands of us.' And he replied, 'Millions will hear, but only thousands will remember.' And we only need thousands. Thank you for giving me a way to fulfill my mission.' [261]

'Following her remarkable NDE,[262] Anne was inspired to pursue various therapeutic trainings and, eventually, to become a practitioner of the *Emotion Code*, when she finally understood what she had seen in her experience and to become one of the thousands needed to carry out this healing work.

'In *Dying to Be Me: My Journey from Cancer, to Near Death to True Healing,* Anita Moorjani describes her four year battle with cancer during which her body eventually became emaciated, riddled with tumors, confined to a wheel chair, unable to breathe without an oxygen tank and, in the last stages, she was taken to hospital in a coma where, it was believed, she would soon die.

'In her moving account, Anita describes how she found herself outside of her comatose body, in an expanded state of awareness with 360 degree perception. She understood the oncologist, beyond the range of normal hearing, 40 feet from her bed, explaining to her husband that, although his wife's heart was still beating 'she's not really in there,' that it's 'too late to save her.' She observed the medics frantically working over her body but noticed she no longer felt any attachment to it. In fact, she felt liberated, with every 'pain, ache, sadness and sorrow' gone. She had never felt so good before.

'She noticed feeling weightless and with a sense of knowing that she could be anywhere at any time and that it felt normal to her, as though returning to a more accurate way of perceiving reality. She heard the doctor saying, 'There's nothing we can do for your wife, Mr. Moorjani. Her organs have already shut down. ... She won't even make it through the night.' She, though, found herself longing to tell her husband that, from her perspective, she was ok, all was ok.

'As her awareness continued to expand, she realized her brother was on a plane, hoping to be in time to say goodbye. She wanted to tell him she was fine. In fact, she had never felt better. Then, it came to her with some surprise that the basic nature of the cosmos is *love.* But why could she now see and understand so much? Who was giving her this information? God? Buddha? Jesus? Suddenly she knew that the divine can be *known directly, as a state of being, that, in fact, she, too, was that state of being.* She also realized that,

'I'm not who I'd always thought I was: Here I am without my body, race, culture, religion, or beliefs . . . Yet I continue to exist! Then what am I? Who

[261] Nelson, B. 2007. *The Emotion Code: How to Release Your Trapped Emotions for Abundant Health, Love and Happiness.* Wellness Unmasked Publishing; pp. 268–274.

[262] Described, by her, also on youtube, https://www.youtube.com/watch?v=LOhCdflip10

am I? I certainly don't feel reduced or smaller in any way. On the contrary, I haven't ever been this huge, this powerful, or this all-encompassing. Wow, I've never, ever felt this way! . . . I felt eternal, as if I'd always existed and always would without beginning or end. I was filled with the knowledge that I was simply magnificent!' [263]

'Then, she found herself wondering why she had always been *so hard on herself?* Because, she now experienced, at a very deep level, that she was loved unconditionally. Why, then, her constant need to measure up and to win others' approval? Why had she not loved herself unconditionally and spoken her truth? Why did she not have this understanding when 'in body?' Why had she never realized that we're not supposed to be so hard on ourselves? She now found herself immersed in a sea of unconditional love and acceptance.

'I became aware that we're all connected. This was not only every person and living creature, but the interwoven unification felt as though it were expanding outward to include everything in the universe – every human, animal, plant, insect, mountain, sea, inanimate object, and the cosmos. *I realized that the entire universe is alive and infused with consciousness, encompassing all of life and nature.* Everything belongs to an infinite Whole. I was intricately, inseparably enmeshed with all of life. We're all facets of that unity – we are *all* One, and each of us has an effect on the collective Whole. . . . [264]

'Anita then faced a choice. She could continue into the spiritual worlds and the cause of death would be organ failure. Or, she experienced her father, already passed on, communicating with her that, she could return to her body and recover. As many other NDErs have experienced, she found the thought of returning to her body very uninviting. But, if she did, she would be well.

'One of the things which is so inspiring about this particular NDE is her subsequently extremely rapid recovery which is medically extraordinary. You should read it.'

'Okay, that's another intriguing account,' said Alisha, 'but, touching as these stories may be, they're all anecdotes. They have no scientific merit, do they?'

'Well, there are *patterns* here, especially if you review the modern NDE literature in more depth, as Chris Carter does in *Science and the Near Death Experience* and Raymond Moody in *Life After Life*. Is it fair to dismiss people's experiences just because we don't have personal access to them?'

---

[263] Ibid, p. 69.

[264] Ibid, pp. 69–70, emphasis added.

'But are people's private spiritual experiences of any public relevance. In any case, how do they relate to our conversations about *macro-evolution?'*

'I mention them because our science currently treats natural history as a wholly materialistic process. Mainstream western science has rejected Cartesian dualism and all concepts of *Spirit, Life Forces and Soul.*

'Within western science, to be a 'vitalist' or a 'spiritualist,' today, is considered unscientific. But the modern evidence from NDE's, OBEs and other psychic phenomena falsifies materialism. It supports the classical hypothesis that there is *more* to us than atomic billiard ball matter and that no purely materialistic theory of Life, be it Darwin's or anyone else's, will be adequate to give a full account of reality.

'As to the public relevance of people's private psychic experiences there is now so much evidence and research relating to NDEs and other kinds of OBEs that it's blinkered just to ignore it. As this researcher writes,

> 'A college professor ... told of dying on the operating table and then instantaneously finding himself walking down the gentle slope of a bright greensward. No one else was there. Nothing else happened. Yet to hear him speak of it, it was as if the greatest of miracles had happened to him, and he now *knew,* he absolutely *knew,* that there was life after death. This brief incident completely transformed his life. Even today he lights up describing the incredible aliveness of the green grass he once walked upon on The Other Side of death.'
>
> 'She continues, 'I have spoken with many people who have described 'the living dark' that [first] greeted them with words like 'soft velvety blackness' and 'warm inviting blanket'. ... these people felt awed by the wonderment of a blackness that appeared *intelligent,* emoted feelings and instilled in the experiencer a sense of peace and acceptance. . . .
>
> '[Others] encountered light . . . They claim that the radiant brilliance of this special light does not blind or burn; it simply accepts, embraces and loves. A sense of worthiness can remain afterwards, changing forever how the experiencer regards him or herself.' [265]

'NDEs and OBEs are individual experiences, Ali, it's true, but they are of tremendous interest to us all...'

'Okay...'

'Because they show us that mind and awareness are not merely the *unintended byproducts of material bodies, which is the current 'scientific' view,* but rather are *transmitted* by them, just as radios transmit their shows from

---

[265] Atwater, P.M.H. 1994. *Beyond the Light: Near Death Experiences.* Thorsons; p. 28, emphasis added.

elsewhere. Often, too, those who have had these kinds of experiences are greatly changed by them. They're more loving and more at peace. They no longer fear death. Surely it's a great gain for our culture to become more aware of these important things?'

'I suppose it would reduce our collective fear of death. But are these people really dead?' Wondered Alisha.

'Yes, in some cases they are clinically dead.'[266]

'But isn't there a difference between *objective* and *subjective* experiences?'

'It depends on what you mean by objective. A tree is not made objective just because we can all see it. We *all* experience the objective tree *subjectively*.

'Currently, our science endorses our collective *subjective* experiences of the tree and calls them objective. But, it is not prepared, yet, to consider the possible truth of numerous individual experiences, such as NDEs and OBEs, even when they share many commonalities and can sometimes be validated. [267]

'No one's experience, even of a tree which we can all see, is fully accessible to anyone else, yet, in normal reality, *we take account* of other people's experiences. We often rely on just one or two witnesses to convict someone of a serious crime. Yet, even though many people, all over the world, have testified to various religious, psychic, mystical and other spiritual experiences and even though their experiences share many commonalities and can, at times, be corroborated and validated, they are often dismissed.

'Unfortunately, as Chris Carter writes, in *Science and the Near Death Experience,* materialism "cannot accommodate corroborated reports" that human consciousness can operate outside of a physical body.[268] And *many of those who think themselves 'scientific' simply refuse to accept any evidence which challenges materialism,* including that for OBEs and telepathically transmitted messages from those passed on.[269]'

'Why are they so dogmatic about this?' Asked Alisha.

'Because, they mistakenly equate science, which is about finding out what is *true,* with materialism, which is a *belief* system. A belief system which may or may not be true. In fact, *'When many scientists and philosophers are confronted with the evidence, their reaction is often anything but rational.'* Carter quotes philosopher of science, Neil Grossman, who stated,

---

[266] Lommel, P. 2010. *Consciousness Beyond Life, The Science of the Near Death Experience.* Harper One.

[267] Carter, C. 2010. *Science and the Near Death Experience.* Inner Traditions.

[268] Ibid, p.235.

[269] Schwartz, G.E. Phd. W.L. Simon. 2002. *The Afterlife Experiments: Breakthrough Scientific Evidence of Life After Death.* Atria Books.

'I was devouring everything on the near-death experience I could get my hands on, and eager to share what I was discovering with colleagues. *It was unbelievable to me how dismissive they were of the evidence.* "Drug-induced hallucinations," "last gasp of a dying brain," and "people see what they want to see" were some of the commonly used phrases. One conversation in particular caused me to see more clearly *the fundamental irrationality of academics with respect to the evidence against materialism.*

'I asked, "What about people who accurately report the details of their operation?"

'"Oh," came the reply, "they probably just subconsciously heard the conversation in the operating room, and their brain subconsciously transposed the audio information into a visual format."

'"Well," I responded, "what about the cases where people report *veridical* perception of events remote from their body?"

'"Oh, that's just a coincidence or a lucky guess."

'Exasperated, I asked, "What will it take, short of having a near-death experience yourself, to convince you that it's real?"

'Very nonchalantly, without batting an eye, the response was: "Even if I were to have a near-death experience myself, I would conclude that I was hallucinating, rather than believe that my mind can exist independently of my brain."' [270]

'As Carter says, for many today, materialism is not merely a hypothesis about reality which is open to "being proved false." No. For many people it has become *"an article of faith that "must" be true, regardless of evidence to the contrary."* [271]

'Yet, as he also notes, it is the *"implicit equation of materialism with science,* that explains the widespread practice of ignoring and dismissing the objectionable evidence [like that for telepathy, OBEs or for intelligent design] as somehow 'unscientific'." [272] Carter, quoting Grossman again,

'Science is a methodological process of discovering truths about reality. Insofar as science is an objective process of discovery, it is, and must be, metaphysically *neutral.* **Insofar as science is not metaphysically neutral, but instead weds itself to a particular metaphysical theory, such as materialism, it cannot be an objective process for discovery.** There is much confusion on this point, because *many people equate science with materialist*

---

[270] Grossman, N. 2002. *Who's Afraid of Life After Death?* p.8

[271] Carter, C. 2010. *Science and the Near Death Experience.* Inner Traditions; p.237.

[272] Ibid, emphasis added.

metaphysics, and phenomena that fall outside the scope of such metaphysics, and hence cannot be explained in physical terms, are called 'unscientific'.

'This is a most unfortunate usage of the term. For if souls and spirits [or intelligent design] are in fact a part of reality, and science is conceived epistemologically as a system of investigation of reality, then there is *no reason why science cannot devise appropriate methods to investigate souls and spirits [or the evidence for intelligent design].*' [273]

'Eben Alexander's bestseller, *Proof of Heaven: A Neurosurgeon's Journey Into the Afterlife* describes his NDE. Especially interesting is that it happened to a previously materialist thinking neurosurgeon who believed mind is produced by the brain and cannot exist apart from it. His experience changed his mind. Given that, by now, there are hundreds of such accounts, is it right to dismiss *every single one* of them as hoaxes or hallucinations, as Michael Shermer sets out to do in a dismissive article in *Scientific American?*[274]

'Isn't it arrogant to tell people that their experiences are not what they think they are just because we don't have personal access to them or because they don't mesh with our world view?

'In *Consciousness Beyond Life, The Science of the Near Death Experience,* cardiologist Pim van Lommel describes how, over twenty years, he interviewed hundreds of heart patients who had died, some for five minutes or longer, before being resuscitated. Of these a significant percentage reported ongoing experience after the medical equipment reported them dead. Some recalled witnessing the actions of hospital staff from out of body perspectives.

'Here's another interesting account. Dr Larry Dossey describes how, during a gallbladder operation, a woman experienced cardiac arrest and died. After she was resuscitated, she was able to describe some of the events taking place around her form, while she was out-of-body, in vivid detail. Intriguingly, she noticed that the anesthesiologist was wearing mismatched socks. What makes this account especially interesting is that Sarah had been blind since birth.[275]

'In *Mindsight: Near-Death and Out-of-Body Experiences in the Blind,* Kenneth Ring and Sharon Cooper investigate the evidence from out of body and near death experiences that the unsighted, including from birth, experience accurate vision during NDEs and OBEs. Ring and Cooper call this kind of out-of-body seeing 'mindsight.' A seeing providing accurate detail, sometimes

[273] Grossman, N. 2002. *Who's Afraid of Life After Death?* p.10–12; emphasis and text in square brackets added.

[274] Did a neurosurgeon go to heaven? Why a Near-Death Experience Isn't Proof of Heaven. M. Shermer. *Scientific American.* April 13, 2013.

[275] Dossey, L. 1997. *Recovering the Soul: A Scientific and Spiritual Search.* Bantam.

from many angles at once, 360°, and often accompanied by synesthetic qualities of multi-sensory knowing. They note,

> 'In general, blind people report the same kinds of visual impressions as sighted persons do in describing NDEs and OBEs. For example, ten of our twenty-one [blind] NDErs said they had some kind of vision of their physical body, and seven of our ten [blind] OBErs said likewise. Occasionally, there are other this-worldly perceptions as well, such as seeing a medical team at work on one's body or seeing various features of the room or surroundings where one's physical body was. Otherworldly perceptions abound also, and seem to take the form characteristic for transcendental NDEs of sighted persons – radiant light, otherworldly landscapes, angels or religious figures, deceased relatives, and so forth.' [276]

'There is, now, decades worth of research into OBEs of various kinds, communications from those passed on and reincarnation.[277] Research which shows that there are *dimensions beyond the physical* where those passed on continue to live and to evolve, including possibilities for reincarnation.[278]

'Typically, though, these findings, conducted by researchers at bodies like the British and American Societies for Psychical Research, the International Association for Near Death Studies and the Monroe Institute, the Afterlife experiments of Dr Schwartz, the medically focused researches of neuro-psychiatrist Peter Fenwick and cardiologist Pim van Lommel, demonstrating that out of body and afterlife phenomena are real, have been ignored.'

'Why? The evidence is interesting at the least.' Queried Alisha.

'Because it conflicts with materialism.'

'But surely once a hypothesis, like materialism, is refuted, it's refuted?'

'In theory, yes. But many people, once they have made up their minds about something, find it difficult to reconsider.'

'Antony Flew, the famous atheist you mentioned earlier, managed it.'

'Yes, but it can be difficult to admit we may have been mistaken, even if, by doing so, existence becomes so much broader, more meaningful and interesting to explore. After all, in a *multi-dimensional* cosmos of intelligence

---

[276] Ring, K. S. Cooper. 2008. *Mindsight: Near-Death and Out-of-Body Experiences in the Blind* iUniverse; p. 75.

[277] Stevenson, I. 1980. *Twenty Cases Suggestive of Reincarnation: Second Edition, Revised and Enlarged.* University of Virginia Press. Stemman, R. 2012. *The Big Book of Reincarnation: Examining the Evidence that We Have All Lived Before.* Hierophant Publishing.

[278] Newton, M. 1994. *Journey of Souls: Case Studies of Life Between Lives.* Lewellyn Publications. Over a thirty year period Newton hypnotically regressed 7,000 clients, building up a picture of the inter-life as his clients, in hypnotic regression, reported on the same themes again and again: including assessment of the completed lives, further experiences in the spiritual dimensions and eventual return to physical life.

and meaning, there are other dimensions and planes of existence to investigate, in addition to the ones we are presently familiar with. And, just as we explore the 'tails' side of the holistic coin of reality – the physical world – using our physical bodies, so, some say, it is possible to enter other dimensions, by using our subtle inner bodies, as the world's shamans and some yogis are able to do.

'In *Less Incomplete: A Guide to Experiencing the Human Condition Beyond the Physical Body,* Sandie Gustus asks:

'How many of us have had an experience that suggests a deeper, unseen reality...? ...déjà vu, intuition, synchronicity, premonition or telepathy; felt an instant sense of recognition or familiarity with a complete stranger ...

'Despite this, most people have no direct experience that we live in a *multidimensional environment* that extends far beyond the boundaries of our physical world, and that *we are, in fact, much more than just our physical bodies.* Fortunately, there is one phenomenon that is natural to all humans that allows us to verify for ourselves, from first-hand experience, that we are capable of acting entirely independently of the physical body in a nonphysical dimension ... The out of body experience (OBE).

'Anyone who has had a fully lucid OBE, and I count myself among them, will tell you that if you can be lucid outside the body, you will find that all your mental faculties are fully functioning, that you can make decisions, exercise your free will, access your memory, think with a level of clarity that sometimes exceeds your usual capacity, and even capture information from the physical dimension that can later be *corroborated* . . . and that **to experience all this provides you with irrefutable evidence that the physical body is merely a temporary 'house'** through which your consciousness (i.e. your soul or spirit) manifests in the physical world. [279]

We walked on through the quiet meadows.

---

[279] Gustus, S. 2011. *Less Incomplete: A Guide to Experiencing the Human Condition Beyond the Physical Body.* O Books. Emphasis added.

25 Proof of Heaven

'When particle physicists report their findings, Ali, we believe what they say. Yet most of us have *no direct way* of testing their truth claims any more than we can those of a shaman or prophet, an advanced yogi, psychic or mystic.'

'But isn't it a case of, 'If you studied hard enough, you too would be able to access this information, unlike psychic or spiritual knowledge which you just have to take on trust, as take-it-or-leave-it revelation?' Commented Alisha.

'That's a typical comment, but it's not true.'

'Ok...'

'Because, just as with particle physics, if we have the interest and the passion, we can *train* in extrasensory perception and learn how to discover other dimensions of reality for ourselves. As the theosophists Annie Besant and Charles Leadbeater demonstrated, it is even possible to study particle physics by means of extra-sensory perception or ESP.'

'Analyzing twenty-two diagrams of the hundred or so chemical atoms described in *Occult Chemistry*... Phillips found it hard to avoid the conclusion that "Besant and Leadbeater did truly observe quarks using ESP some 70 years before physicists proposed their existence". [280]

'Yes, the majority of us don't have the time, enthusiasm or the native talent to train in such techniques any more than we have the time, passion or talent to be particle physicists, brain surgeons or astronauts. But, in theory, we could.

'Not everyone knows this, but since the middle of the nineteenth century there has been a lot of research showing that gifted psychics can provide good evidence not only for ESP but also for Life after death. For example, in *Science and the After Life Experience: Evidence for the Immortality of Consciousness,* Chris Carter explores the extensive *modern, non-religious, evidence* for Life after death. He discusses 125 years of studies by independent researchers and the British and American Societies for Psychical Research which firmly rules out hoaxes and hallucinations. Studies which show – beyond reasonable doubt – that the afterlife is real.

'One well-known and very long-running study was called the *cross-correspondences.* It was initiated by Frederick Myers who was a Cambridge classics scholar, a brilliant man whose transmission theory of the mind–brain

---

[280] Tompkins, P. 1997. *The Secret Life of Nature,* Thorsons, pp. 68–69.

relationship was taken up and developed further by his friend, the famous author, Williams James.[281]

'Myers was also one of the founders of the British Society for Psychical Research (BSPR). He wanted to show, beyond reasonable doubt, that information transmitted through spiritualist mediums was not merely the result of an unusual ability to read the minds of surviving relations sitting for spiritualist readings, an idea called super ESP, but evidence for life after death.

'He proposed that, after he had passed on, he would attempt to send a series of telepathic messages to different mediums in different parts of the world, messages which, on their own, would be meaningless but, if put together, would make sense. He and his colleagues at the BSPR believed this would provide good evidence of survival rather than for super ESP.

'After Myers passed on in 1901 more than a dozen mediums, in different countries, began receiving a series of incomplete scripts which they channelled through automatic writing. The pieces were signed 'Frederick Myers.' The scripts contained obscure material from and references to the classics which made no sense on their own. The messages asked the mediums to contact a central address where the scripts were assembled and, then, made sense.

'Later, Myer's colleagues at the BSPR, Professor Henry Sidgwick and Edmund Gurney, also transmitted scripts after passing on. More than 3000 scripts were transmitted over 30 years, some more than 40 pages long. In total, 12,000 pages, in 24 volumes. Later, Colin Brookes–Smith patiently studied this material and stated, in the *Journal of the BSPR*, that **after-death survival should now be regarded as sufficiently well-established to be beyond denial by any reasonable person**. [282]

'In *Science and the After Life Experience* Chris Carter investigates a variety of historic and contemporary accounts of past-life memories, visits from those passed on, telepathic communications via mediums and automatic writing, as well as the scientific methods used to confirm these experiences. Notably, in examining the evidence for the spiritual phenomena he discusses, he carefully explores possible alternative explanations. Only when these have been reasonably ruled out does he conclude that the phenomena are what they seem to be. Dr. Pim van Lommel, author of *Consciousness Beyond Life,* writes: 'The evidence in favor of an afterlife is vast and varied. . . .we do indeed have

'strongly *repeatable evidence* for the continuity of consciousness after physical death . . . What all these cases show is that ... human consciousness

[281] Kelly, Kelly, Crabtree & Gauld 2009 *Irreducible Mind: Toward a Psychology for the 21st Century.* Rowman and Littlefield.

[282] Zammit, victorzammit.com

can exist independently of a functioning brain. When one has read the *overwhelming evidence* as described in this excellent book, it seems quite impossible not to be convinced that there should be some form of life after death. *Any continuing opposition to the evidence is based on nothing more than willful ignorance or ideology.'* [283]

'Dr Lommel is right. You see, Ali, it is not good quality, non-religious, evidence for spiritual phenomena which is lacking but the widespread equation of science with *materialism* which prevents its mainstream scientific acceptance. One thing which amuses me, though, is how materialist skeptics of the supernatural always demand ever more proof for it, yet no evidence, no matter how good or well corroborated, is ever good enough for them. Yet these same skeptics are quite untroubled by the lack of convincing proof for Darwin's theory, let alone for the claim that *unintelligent chance* is incredibly creative and could be the cause of their own intelligent minds.'

'Dr. Larry Dossey, author of *The Power of Premonitions* and *Healing Words,* is more scathing than Dr. Lommel, he writes:

> 'Chris Carter's *Science and the Afterlife Experience* ... is a withering rebuttal of the perennial, timeworn, anemic arguments of skeptics. This book is extraordinarily important, for, as Jung said, 'The decisive question for man is: Is he related to something infinite or not? That is the telling question of his life.' This brilliant book is an antidote to the fear of death and annihilation. It will help any reader find greater meaning, hope, and fulfillment in life. [284]

'Earlier than Carter, some distinguished, but initially highly skeptical, scientists also investigated telepathically transmitted, afterlife communications – often with a view to debunking them. Yet, eventually, they all concluded that life and mind beyond the body were facts. Sir William Barrett F.R.S. (1844–1925), Professor of physics at the Royal College of Science in Dublin, reported that he was, "absolutely convinced of the fact that those who once lived on earth can and do communicate with us. It is hardly possible to convey . . . the strength and cumulative force of the evidence." [285]

'Sir Oliver Lodge F.R.S. (1851–1940), world famous for his pioneering work in electricity and radio and for developing the spark plug, concluded that *"survival is scientifically proved by scientific investigation."* Professor Camille

---

[283] Carter, C. 2012. *Science and the Afterlife Experience.* Inner Traditions, review, amazon.

[284] Ibid.

[285] This and the following quotes are from Michael E. Tymn, author of *The Afterlife Revealed,* White Crow Books.

Flammarion (1842–1925), founder of the French Astronomical Society, investigated psychic phenomena for decades and came to the same conclusion.

'Professor James J. Mapes (1806–1866), an expert in chemistry, set out in the 1850s to rescue his friends involved with the then spiritual mediumship craze. Yet, after investigating many mediums he changed his views and stated,

> "The manifestations . . . are so conclusive [that they establish]: First, that there is a future state of existence, which is but a continuation of our present state of being . . . Second, that the great aim of nature, as shown through a great variety of spiritual existences is progression, extending beyond the limits of this mundane sphere . . ."

'These eminent scientists and thinkers, Ali, were all cautious and often highly skeptical when they began their investigations. It was only after a lot of research that they accepted the evidence. Many of them were highly practical people who made major, sometimes world-changing, discoveries in other fields. They were free thinkers who braved opposition not only from materialists but also from the traditional religious authorities.

'But why would the religious authorities oppose the kind of modern, *non-religious,* evidence you are referring to, such as that from NDEs and for Life after death? Surely this kind of evidence would strengthen their own ancient claims for a multi-dimensional cosmos of intelligence and meaning?'

'Because, Ali, some of the contemporary evidence for the afterlife can be quite challenging to long held religious views. For example, ancient, but outdated, ideas about 'heaven and hell'. The modern evidence, from NDEs and OBEs and from telepathically channelled communications from those passed on, being that, in some respects, Life after death is not too dissimilar to the Life we already know. We still have bodies – at different vibrational frequencies from the 'bio-suits' we use here – and it is our attitudes and states of mind which govern where we find ourselves in the after-life realms, not some irrational scheme of eternal punishment or boring, over-pious heaven.

'Another thing from modern psychic research, which conflicts with many traditional religious teachings, is that, telepathic messages from the afterlife tell us repeatedly, it's not so much the religious creed you adhered to, or whether you had one at all, but how you *were* and how you *acted* during your time on the planet, which is a measure of your spiritual progress in a given lifetime.

"Did you seek the good, the beautiful and the true? Were kindness and the golden rule your guides?' Not, 'Did you *believe and do* precisely everything you were told to believe and do by your particular religious tradition."

'If only we knew more.' Alisha said wistfully.

'Well, we are gradually learning more about the afterlife. But, right now, our science does not accept nor is it interested in the idea that we exist within a *multi-dimensional* cosmic system. Right now it maintains that only the physical worlds we know are real.'

'But what could be more interesting than whether there is anything beyond our 'solid' physical world itself consisting, ironically, almost entirely of S...P...A...C...E? 9. .9 - 9 . .9 . .9. . .9. .9. .9. .9. .9. .9. .9. .9. . .9. . . 9. . .% of it, said Alisha.

'Yes, Ali, I agree. We like to think of ourselves as an inquiring and openminded society, yet our science culture today is remarkable incurious about these things. Despite the fact that there are many valid ways to test the classical hypothesis that mind continues beyond the body and to demonstrate that the spirit and the soul are real, few seem interested.

'Unfortunately, many so called skeptics of the non-physical are close minded. The evidence for all kinds of psychic and supernatural phenomena, including that for mind beyond the body, collected over many decades now, doesn't match their philosophy, so they either ignore it, allege fakery or demand ever 'more proof,' even as they refuse to study the, by now, extensive and good quality evidence which they themselves call for.

'As this geophysicist Ph.D. and reviewer of *Randi's Prize: What Sceptics Say About the Paranormal, Why They Are Wrong and Why It Matters* notes,

> '[Psychic phenomena] have been studied in painstaking detail for 150 years by highly qualified scientists, including, in recent decades, *some of the most carefully executed scientific experiments ever conducted* with multi-layered experimental controls that put other fields of science to shame.
>
> 'Because researchers in this field are under unrelenting, often vicious assault, they control even against absurdly improbable and unrealistic forms of cheating and fraud among other things, problems that most scientists don't have to think about at all . . . *Statistically and taken as a vast body of work, their results are rock solid* . . .
>
> 'The scientific facts are in; they're well-proven and extensively documented – many tens of thousands of pages of detailed studies.[286] *The demand for more proof is simply a ploy.* ' [287]

We walked on in the darkening hills, the birds and the cattle settling down for the night, the river quietly flowing by our side.

---

[286] Radin, D. 2009. *The Conscious Universe: The Scientific Truth of Psychic Phenomena.* Harper One.

[287] McLuhan, R. 2010. *Randi's Prize: What Sceptics Say About the Paranormal, Why They Are Wrong, and Why It Matters.* Matador. Amazon review. by Geophysics Ph.D, Sun Dog.

26 The Great Mystery

'Our amazing world, Ali, and the star spangled cosmos beyond, all come from something, be it mindless or intelligent. But, if intelligent, it doesn't force itself upon us. We are free to think and to believe as we wish – whether or not our ideas are true and and whether or not they correspond accurately with reality.'

'You mean free will?'

'Yes, the gift, or the challenge, we call free will. But, some say, if we are prepared to use our free will to still our busy minds, to meditate, to visualize or to pray, we may find that the *final Source and Uncaused Cause* from whose *Shimmering Living Fields* we emerge into this, our temporary, reality, whispers to us in subtle ways which tell us *who and what* we really are.

'Not, after all, orphan children of a heartless reality which doesn't care for us, because it cannot care, but the precious offspring of an *amazing final Source and Uncaused Cause* which is not only intelligent but also loving. Not, after all, a Mindless, Unintelligent, Accident (MUA), but an Amazing, Ultimate Miracle or AUM. A mighty AUM expressing, cosmically, as a trinity of mind-intelligence-creativity, love-feeling-wisdom and will-purpose-action, hologramatically reflected in us in head, heart, belly and limbs. A mysterious cosmic AUM [288] in whose image we, as Beings of consciousness, love, purpose and creativity are made.'

'You are going all mystical on me...' said Alisha, smiling gently.

'You could put it like that, but existence *is* both a miracle and a mystery isn't? An *impossible* mystery, if you think about it.

'Why *'impossible'?'*

'Because how come there is anything at all, rather than nothing? Even just one atom? Isn't it easier to imagine there being no-thing, rather than some-thing? Why should there be any *thing,* even a quark? No one can say.

'Yet, it *is* possible. But is this 'possible' intelligent or meaningless? Does reality emerge from *a meaningful* final Source and Uncaused Cause which *just Is,* and to which humanity gives its traditional religious names, or mindless?

'Darwin and his companions in the materialist faith choose mindless. Why? Because the subtle sources of reality cannot be physically seen, they can only

---

[288] In Hinduism, the sacred syllables A-U-M represent the primal sound and vibration of God or Supremely Intelligent Being giving creative expression to the universe, from no-thing to some-thing.

be intuited, logically inferred or known, to some extent, by extrasensory and spiritual means. This causes many to doubt that there may be anything *more*.

'This is why famous materialists like Francis Crick tell us that,

> 'You, your joys and your sorrows, your memories and free will, are in fact no more than the behavior of a vast assembly of nerve cells and their associated molecules. As Lewis Carroll's Alice might have phrased it: "You're nothing but a pack of neurons."[289]

'But this kind of extreme reductionism can give us no convincing ideas as to what *Soul* states like Awareness, Subjectivity, the ability to Experience, or Will, Desire and Love, Intelligence and Creativity really are. It can tell us nothing about the inner *Soul and Spirit* of things. Without troubling to think terribly deeply, it assumes them to be mere accidents of matter, and, there, 'intellectually satisfied' with this thin fare, its curiosity and its inquiry ends.'

'That does seem very limited...' Commented Alisha.

'It is. How, for example, can anyone seriously believe, in the name of rational science, that tiny, DNA CODED Cells *of an almost unbelievable complexity* are mere chemical flukes? Nothing special? Butterflies and Oak Trees, Eagles and Bears, Monkeys and People, all mere chemical accidents?

'If these fascinating things are *not* to be understood as amazing expressions of *intelligence, beauty and Soul* made visible, then, it seems to me, we no longer know what these words *mean* and important parts of our science are no longer instruments for truth and understanding but irrational vehicles for peculiar distortions which, far from enlightening us, darken our understanding of *who and what* we truly are, and of multi-dimensional nature all around us.

'I find it hard to understand such thinking, Ali. When some say, as Darwin did, that life's difficulties and nature's harshnesses make them doubt whether there can be any positive, ultimate, Higher Power or Loving Final Force of the kind to which humanity gives its traditional religious labels I can understand.'

'So how do you answer their point?'

'Well, the question as to whether something, like a poisonous spider or a dangerous virus, has an intelligent source is quite apart from the consequences of its being. We design many dubious things, but no one denies the *intelligence and purpose* which go into their creation. As to the problems of natural evil and human suffering, the world's religions have some ideas or theodicies which may help, even if they cannot answer all our questionings.'

'Such as?'

---

[289] Crick F. 1995. *The Astonishing Hypothesis.* Scribner, p. 3, text in square brackets added.

'Well, one idea is that without opposites or relativities – hot and cold, left and right, up and down, pleasant and less pleasant – there could not *be* manifested existence. There could only be undifferentiated oneness and stillness. Another is that a mysterious spiritual falling away took place, aeons ago, on subtle spiritual levels, and led to this beautiful but challenging world.

'One aspect of this idea is that there are *subtle spiritual forces,* sometimes conceptualized as fallen angels or demons, which subvert things on subtle, *non-physical levels* by creating some of the very difficult conditions – including negativity and disease – with which we have to battle. These challenging forces, while difficult to understand, do, nonetheless, play their parts in creating the great panoply of existence.

'Another idea is that our world is a school for Souls, with tough rules. Others say that 'free will' means free will, otherwise we would just be *puppets.* This is why Christ's teachings were so radical. They still are. He taught that we are not robotic automatons obliged to return violence for violence every time we are wronged. He said we could use our *free will* in a different way.

'Others mention that beyond our sufferings in this world there's an ultimate goodness to reality, a tremendous light and love behind all outer appearances.

'Many of those who have been through near death and other kinds of out of body experiences report that they experienced spiritual dimensions which were *intrinsically intelligent, and amazingly loving.* But, whatever the correct answers to the mysteries of earthly darkness and suffering, the materialist claim that living nature, and us within it all, is just a long series of mindless chemical coincidences is not only impossible, it simply doesn't make sense.

'If we make something clever, Ali, we consider it to embody purpose and intelligence. But if nature expresses something vastly more ingenious in its *dynamic, self-repairing, self-reproducing functionality and far, far more intricate* than anything we can make, we are to follow Darwin and persuade ourselves it's just a lucky chemical coincidence, without purpose or meaning?

'A few chemicals + mindless natural laws + lots of random swilling around are all that's needed? When was this ever a genuinely scientific way of looking at reality? Or was it always just a rather inelegant way to replace the old creation stories which were never intended to be taken literally, except, perhaps, by the most simple of folk? No modern thinker, discussing intelligent design (ID) versus Darwin's theory of unintelligent design (UD) is claiming that some Sunday school, Father Christmas type god is sitting on some far off cosmic cloud designing Life Forms.

'No, all they are saying is that we and today's scientists are perfectly well-equipped enough to work out whether we are observing intelligent and meaningful phenomena or purely random. But, if we take Darwin's materialist line and deny all intelligence and purpose in Nature, then, logically, we have to deny all intelligence and purpose in ourselves.'

'Because we emerge from Nature and must share its qualities?' Mused Ali.

'Yes, and, then, everything *we* say becomes meaningless, not least the words of those famous materialists who, so strangely, use their spiritually-derived meaning making abilities to make meaning by *denying* meaning, their *inborn, spiritual intelligences to deny all intelligence* in the amazing cosmic matrix which gives birth to them.

'We use our intelligence – not randomly tumbled about parts – to make Jaguar Cars and Jaguar Jets. These clever creations of ours, although they are like wooden spoons compared to their living, feline inspirations, are designed to evoke, in a minor way, the keen intelligence, the predatory purposefulness and the stunning speed of the big cats of South America.

'There is in fact now a whole new science called *biomimetics.* It is devoted to imitating systems in nature for the purpose of solving complex engineering and other design problems. Yet, despite this, we continue to deny all intelligence in this same *amazing nature* which gives birth to us?, – whose un-admitted designs we copy! We do our best to *intelligently* learn from her but *her* designs are not designs, and, unlike ours, they are not clever at all?

'They're just aimless, dumb flukes which we, who *are* clever, can base our own intelligent designs upon! We put millions of hours into designing AI driven robots but mindless chemical coincidences could give rise to us?

'Irrational! This way of thinking is not only illogical, Ali, it is degrading to us and damaging to our science. A narrow, reality-shrinking worldview driven, in part, by pseudo-skepticism and ego, not wisdom or common sense.

'In his book *The Science Delusion: Freeing the Spirit of Inquiry,* biologist Rupert Sheldrake explores the now widespread delusion that our "science already understands the nature of reality, the fundamentals are known and only the details remain to be filled in." He argues that science is now,

> 'being constricted by assumptions that have hardened into dogmas. The 'scientific worldview' has become a *belief system* [rather than open minded inquiry]. All reality is *material* or physical [and all evidence showing otherwise is ignored or denied]. The world is a *machine* [which unlike all other machines is a mere accident], made up of dead matter [vibrating at trillions of cycles per second]. Nature *is purposeless* [the Lion hunts for no reason, Daffodils grow for no reason, Caterpillars transform into beautiful Butterflies for no reason, Bach composed preludes and fugues for no reason, Picasso painted for no reason, Einstein discovered that E=mc2 for no reason].

'Consciousness is nothing but the physical activity of the [accidental and purposeless] brain. ['We're all zombies. Nobody is conscious.' Dennett.] Free will is an illusion ['You're nothing but a pack of neurons,' Crick.] God exists only as an idea in human minds imprisoned within our skulls [and cannot be a direct spiritual *experience*[290]].

'Sheldrake examines these dogmas, and shows, persuasively, that science would be better off without them: freer, more interesting, and more fun. In *The God Delusion* Richard Dawkins used science to bash God, but here Rupert Sheldrake shows that Dawkins' understanding of what science can do is old-fashioned and [is] itself a delusion.'[291]

'In *The Guardian,* Mary Midgley commented:

'We must somehow find different, more realistic ways of understanding human beings – and indeed other animals – as the active wholes that they are, rather than pretending to see them as meaningless consignments of chemicals. Rupert Sheldrake, who has long called for this development ... shows how *materialism* has gradually hardened into a kind of anti-Christian [anti-spiritual] principle, claiming authority to dictate theories and to *veto* inquiries on topics that don't suit it, such as unorthodox medicine, let alone religion. ...

'The 'science delusion' of his title is the current popular confidence in certain fixed assumptions – the exaltation of today's science . . . as a final, infallible oracle preaching *a crude kind of materialism* . . . His insistence on the need to attend to possible wider ways of thinking is surely right.'[292]

'Thankfully, Alisha, gradually, more people are exploring wider ways of thinking. For example, Chris Carter's trilogy on near death experiences, psychic phenomena and Life after death shows, with numerous examples, that Consciousness and Mind are not, ultimately, matter dependent.

'Now, all through our walk I've been arguing that there's an *intelligence* to how living things are made. But we've never attempted to define intelligence.'

'Okay,' said Alisha, looking intrigued, 'how do you define it?'

'Well, intelligence is an *ontological* quality, an intrinsic aspect of *Being, of All that Is*. For the religiously inclined, it is a fundamental attribute of *Spirit or the Divine*. It is not physical but we all recognize it when we see it, be it in the

---

[290] James, W. 1902. *The Varieties of Religious Experience*. Penguin Classics.

[291] Sheldrake, R. 2012. *The Science Delusion*. Coronet; product description, amazon.

[292] Ibid, amazon, Mary Midgley, *The Guardian;* emphasis and text in square brackets added.

mind of an Einstein, the design of an elegant Car, in the beauty of a jaguar Cat or in the 24/7 dynamic functioning of Living things.

'It is not material but it pervades all and it informs all. Without it nothing would work. Think of the ribosomes and proteasomes, the bat's sonar and the bombardier beetle's amazing weaponry, birds' feathers and birds' lungs, the magical transformation of the green caterpillar into the beautiful, multi-hued butterfly, all masterpieces of bio art and bio engineering. Think of the smart-materials functionalities of the atomic elements, Life's complex DNA CODES. *Nature is pervaded by intelligence. She is intelligence made visible.*

'We have no reason to think the spiritual quality we call intelligence randomly arises from unintelligent atoms bumping and clumping into each other, for no reason or purpose at all, before, eventually, becoming human shaped, and, voila, 'intelligence.' The very primordial smartness of those atoms, each a thoroughly teleological mechanism in its own right, is an expression of the same *amazing, living, dynamic, intelligence* which our own astonishingly complex bodies and amazing minds arise from.

'We recognize intelligence, too, as an inherent ability to respond to reality with awareness, evaluation and skillful action. It is a characteristic we share with animals and even plants.[293] Yet, it is held by many today to be nothing more than an *accidental byproduct* of an accidental brain. This is, today, the mainstream view, the respectable 'scientific' view, but it is not logical.

'We only have to look just a little more closely to see that intelligence is *not* just a quality of our minds, but it is all pervasive. It's a basic quality of the very ontology of reality, as many spiritually realized people, including many out of body experiencers have also found. P. M. Atwater describes NDErs who,

'felt awed by the wonderment of a blackness [or, at times, a light] that appeared intelligent, emoted feelings and instilled in the experiencer a sense of peace and acceptance.'[294]

'Anne Horne commenting on her NDE, in *The Emotion Code,* says,

'I was going back, home, and on my way, there was a light, a tunnel. I felt like I was being pulled by my heart toward a wonderful place. In that moment, I was encompassed by all this innate intelligence and tremendous love. And I just wanted to go home. It was fantastic.' [295]

---

[293] Tompkins, P C. Bird. 1989. *The Secret Life of Plants.Harper* Perennial.

[294] Atwater, P.M.H. 1994. *Beyond the Light: Near Death Experiences.* Thorsons; p. 28.

[295] Nelson, B. 2007. *The Emotion Code: How to Release Your Trapped Emotions for.*

'What, too, of *Spirit* qualities like *beauty, goodness and truth?* The mindless atoms of materialism know nothing of love and goodness or beauty and truth. Yet, we *do* know of them and we all seek our own versions of them.

'But why do you think, Ollie, these *essential* qualities and conditions, like *awareness and sentiency, feelings and desires, will and motivation, love and joy, curiosity and truthfulness, creativity and intelligence,* all exist?'

'No one knows, Ali. All we can ask is this: Does our universe consist in outer forms alone? Or, does it consist, also, in *inner aspects, Life and Soul forces, Subjective aspects, including meaningful, subtle, Spiritual dimensions?*

'It is true that we cannot physically see the *subtle, Spiritual dimensions of the panentheistic vastness* in which we live and move and have our being, nor *how It* gives rise to the outer forms, the Stars, the Planets and the *Soul-Life-In-Forms* of the Plants, Animals and People. Physically, most of us can only see its outer shapes and forms which make up the macrocosmic whole. Yet, with our spiritual eyes, mediated by reason, we can see so much more.

'This is why, for the world's first peoples, its yogis and its sages, its psychics and its mystics, its prophets and its shamans, the beautiful worlds of nature, and the star spangled cosmos beyond, are all part of the great Mystery, *the Great Spiritual Mystery.*

'They are its living manifestations. Yet, if we wish, truly, to connect with the deeper dimensions of things, we must look *within* ourselves, not just outwards with telescopes and downwards with microscopes.

'Remember, Jesus reminded us, *'the Kingdom of Heaven is within.'* He didn't say, 'It's out there in the sky somewhere and, one day, when you have really powerful telescopes, you'll be able to see it.'

'The subtle, *inner* realms of reality, Ali, the other dimensions and vibrations of existence, can, during our physical lives, only be accessed by *inner* practices like meditation, visualization and prayer. We cannot enter them with our purely physical senses or their extensions. Materialist thought, by contrast, denies that any such inner ways of knowing are possible nor, indeed, that there is anything more than the *random dust and random accidents* of its dreary belief system to know.

'In materialism, there is only ever the *outer,* the purely mechanical and the purely materialistic to explore. There are only 'accidental' sensors (eyes and ears) which 'purposelessly' report to the central, but accidental, processor of the staggeringly complex, but 'unintelligently designed,' brain and the apparently resultant Human Being with *wishes and desires, feelings and dreams,* is not truly a living *Subject or Soul* at all, but merely a meaningless Object, an 'it.' A deluded 'it,' mistaken in believing itself to be anything other than an accidental robot, just a pointless "pack of neurons."

'Never mind that beyond materialism's iron curtain of the mind, beyond the sombre walls of Darwinia – where all thought and talk of intelligence and design are academic thought crimes – no robot *ever* falls together by pure higgledy-piggledy, entropy-defying, pointless-random-swilling-about luck.

'Materialist thought cannot, of course, acknowledge the intrinsic intelligence in living things or it would disappear. For the same reason, it cannot acknowledge that there is any true *inner Being or sentient Soul* here to know.

'No, in its strange, de-Souling schemes there can be no *living Souls* who gaze at us from a friend or lover's eyes. There are, as the materialist philosopher Daniel Dennett maintains, just random, bio-robot zombies which may think they have free will, loves and desires but, really, they are deluded.

'This unreal, unintelligently designed human of current scientific thinking has no true subjectivity, no true I-Am-That-I-Am Beingness, it cannot have a real or true Me-You relationship. Why not? Because it's all *object* and no *Subject,* all 'tails' and no 'heads,' all mechanism and no *Soul,* just the mindless solid little billiard ball atoms of the materialist belief system.

'If this random bio-robot of *materialist faith and belief,* in apparently *Soulful* and apparently Human form, says, 'I love you, Ali,' don't believe it, any more than you'd believe any machine because there's *no one* truly there. Nothing there to *feel* and nothing there to *know,* because in the *de-souled and de-rationalized* universe of so called scientific materialism no one is ever truly at home. We're all just 'unconscious zombies,' as the materialist philosopher Daniel Dennett puts it, nothing but a 'pack of neurons' as Francis Crick had it.

'But that is all such nonsense!' Said Alisha, laughing.

'Yes, Ali, I agree, it very definitely is. But it *is* the current orthodoxy. According to this supposedly 'scientific' view we are just cosmic accidents, without purpose or meaning. So called scientific materialism attempts the illogical and the impossible, making science unreasonable and philosophy foolish. Is this why the late Stephen Hawking argued that *philosophy is dead?* Because science has killed it?

'If this were true it would be like arguing that science has killed reason and common sense. Philosophers, he said, 'have not kept up with modern developments in science and... philosophical problems can [now] be answered by science...'[296] Really? No. Brilliant as he was, he was wrong. Philosophy means the *love of wisdom.* But modern science, amazing as it is, is 'technical' knowledge. It is not philosophy. It cannot tell us what is wise or good.

'Unfortunately, materialist thinkers, from Darwin to Dawkins, use their Soulful caring, their love and their desire – all *non-material, Soul and Spirit*

---

[296] Zeitgeist Conference, by Google, 2011. Hawking said that "philosophy is dead." Wikipedia.

qualities – for what is good, beautiful and true, also *Soul qualities,* to argue for total Soullessness, total cosmic unintelligence and total meaninglessness. How illogical!

'Global reality is unintelligent and completely purposeless in their view, at all levels and dimensions. Yet, if there were *no* intelligence or purpose in Nature, Ali, how could we, who emerge from it, be intelligent, make sense of it or do intelligent, purposeful science on it?

'Science is mean't to be a logical and rational activity. How, then, can one advise thoughtful young people to go anywhere near certain areas of this discipline which illogically tells them, in the name of an irrational pseudo-rationality, that they are 'nothing but a pack of neurons,' that 'we are all unconscious,' that there is *no intelligence, teleology (purpose) or design* in nature and all living things are mere accidents?

'Yet, everywhere we look in nature we see purposeful functioning. There's not one atom that's not *doing* something or *functioning,* even if all it is doing is helping to make a rock. Yet, without that purposeful functioning, of all of nature's components, there would be no atom, no rock, or any thing at all.

'Purposeful functioning (teleology) is woven into the very ontology of reality. There can be no movement, which means work and functioning, not even of one atom, without motive force. Because without motive force nothing would ever move. There'd be absolute stillness.

'But every movement, even of an atom, *is, by definition, working or functioning,* which, by definition, implies *purpose,* which, by definition, implies *motivation* which, by definition, implies *will, desire and love.* But without an *intrinsic elegance and intelligence* to all of nature's movements and laws, at all levels and dimensions, nothing would ever work.

'Yet nature's mechanisms generally work very well, which is why reality is indivisibly made up of Mind–Intelligence, Love–Desire, Will–Purpose. It is from this mysteriously inter-weaving trinity of Soul–Spirit qualities, imaged micro-cosmically in us in Head (Mind-Intelligence-Creativity), Heart (Joy-Love-Desire-Curiosity), Belly and limbs (Will-Purpose-Action), that reality emerges.

'Then, putting aside the fact that no robot could ever arise by mere luck – randomly tumbling chemicals or robot parts about will never work – is it true that we have no real consciousness or free will as materialists like Crick and Dennett claim? You see, Ali, no mere robot has *real* consciousness or real love or real desire.'

'Go on...'

'Because no mere robot made out of the apparently mindless rocks and oils we drill out of the ground has any *true, living Awareness,* or any real ability to *freely* choose the good or its opposite.

'Materialist thought, though, cannot seem to grasp, among all the fashionable, unthinking talk about AI and robots taking over the world, that no robot, no matter how apparently clever, will ever have *true* free will because it will have no true *subjectivity or Soul.* It may be made to *look* as though it is conscious but it won't be. It will just be a mindless automaton, made by *us.* A programed, computerized puppet, which we are all too ready to worship, forgetting that it's just an algorithmic idol which *we* have made.

'Yes, a robot may be designed to *look* as though it cares, as though it wants to love or to laugh and to play but it will be a hollow mockery of anything truly *Living or Soulful* because it will have no true 'I-Am-that-I-Am-Being.'

'Behind it will always be the human intelligence which created it, an intelligence which cannot give it *true* free will. For we, unlike *Spirit,* cannot confer *true, living Soul* upon it, but only inanimate sensors and a lifeless CPU.

'Unfortunately, Alisha, too many today don't see that an intelligently designed listening device is not a *real, living, inner Experiencer* who hears. A robot, no matter how apparently clever, will never be truly *Sentient.* No loud-speaker is a *living Subject* who speaks. No light sensor is *a Soul* who perceives. A robot's CPU can only recycle the programs coded into it, by us. A robot, no matter how seemingly clever, cannot *truly* choose. It can only react based on the algorithms programed into it – *by us.*

'At the same time, if the amazing cosmic matrix *we* emerge from had no intelligence neither would we. If it had no purpose neither would we. If it had no love neither would we. This is why *panentheism* makes sense.

'What is panentheism?' Wondered Alisha.

'It's the classical understanding that *a self-arising, self-existing, Cosmic Animating Force,* to which humanity gives its traditional religious names, (God, Brahman, Allah, Buddha Nature the Great Spirit, the Tao), both gives rise to and pervades Nature while, at the same time, timelessly extending beyond it.'

'You mean *god immanent and god transcendent,* to use religious language?'

'Yes. The mysterious *Spirit* which gives rise to all is both *within* all and *beyond* all. Panentheism also reflects the classical insight that we are microcosms of the macrocosm. That we are hologramatic reflections of the whole, imbued with intelligence, love, curiosity, creativity, purpose and will.

'Our challenge, though, is that, unlike robots, we *do* have free will. We are free to choose the good, the beautiful and the true, or, their opposites. We turn towards a *Higher Power,* whatever name we may give it, not merely because our ancestors were afraid of thunder and lightning, as some say, but because something in us hears a deeper call. Something which knows that we are *born of this same Higher Nature* to which humanity gives its traditional religious names.

'We are born from its *Shimmering Living Fields,* we come from It and, ultimately, we return to It. This is why the *Soul* in us loves the truth and seeks the truth. This is why our quests for scientific and philosophical truths, and for the good, the beautiful, and the true, are also part of our True Nature.

'It is this *same, amazing, Living, Intelligence, **in us,*** which pursues science philosophy, art and mathematics, and seeks for the truth of Nature in all her aspects, obvious and subtle, superficial and deep, and, for spiritual seekers, for enlightenment and spiritual fulfillment.

'Something which whispers to us that we are *more* than mere cosmic accidents and that if that's all we were we wouldn't, in any case, care. We came into this world, as Wordsworth so beautifully put it, "trailing clouds of glory." Something inside us remembers and calls us home. Our lives are not merely "tales told by idiots, full of sound and fury, yet signifying nothing," as Shakespeare's tragic Macbeth cried out, but meaningful.

'As Anita Moorjani discovered in her remarkable NDE, we are *spiritual* Beings having physical experiences, not the reverse. We are, as she found, to her astonishment and her delight, magnificent beyond our own imagining, not in some childish, egotistical way, *but in our very essence.*

'It is this *precious, subtle, essential, ground of our ultimate Source* which gives us our own deepest values of the most good, the most beautiful and the most true.'

'Ah, these are deep mysteries...' Alisha sighed.

'Yes . . . they are. Perhaps we can go into them some more one day. But, for now, as our gentle walk by this lovely river in this beautiful valley comes to a close, thank you for giving me a chance to share these ideas with you.'

'You're very welcome, it's been very interesting.'

'I just wish, my dear Ali, that more people today could see that the materialist belief system and all purely materialistic evolutionary thinking, like Darwin's, while currently fashionable, both as science and as philosophies of existence, don't work. Not only this, but they instill nihilism and despair wherever they take root, and within science, especially within science, they block progress because they block a deeper understanding and a more open minded inquiry.'

We headed home.